# The Cuban Cure

# The Cuban Cure

Reason and Resistance in Global Science

S. M. REID-HENRY

The University of Chicago Press
Chicago and London

S. M. Reid-Henry is a lecturer in the Department of Geography at Queen Mary, University of London.

The University of Chicago Press, Chicago 60637
The University of Chicago Press, Ltd., London
© 2010 by The University of Chicago
All rights reserved. Published 2010
Printed in the United States of America

19 18 17 16 15 14 13 12 11 10      1 2 3 4 5

ISBN-13: 978-0-226-70917-8 (cloth)
ISBN-10: 0-226-70917-5 (cloth)

Library of Congress Cataloging-in-Publication Data

Reid-Henry, Simon.
 The Cuban cure : reason and resistance in global science / S. M. Reid-Henry.
   p. cm.
 Includes bibliographical references and index.
 ISBN-13: 978-0-226-70917-8 (cloth: alk. paper)
 ISBN-10: 0-226-70917-5 (cloth: alk. paper)
 1. Biotechnology—Cuba. 2. Science—Cuba—History—20th century.
I. Title.
 TP248.195.C9R44   2010
 660.6097291—dc22
                                                                    2010007676

♾ The paper used in this publication meets the minimum requirements of the American National Standard for Information Sciences—Permanence of Paper for Printed Library Materials, ANSI Z39.48-1992.

*To my parents*

CONTENTS

*Acknowledgments / ix*

INTRODUCTION / 1

ONE / A Biotechnology Story / 13

TWO / Imagining Science / 23

THREE / Making Space for Science / 41

FOUR / Science City / 63

FIVE / Sticky History / 89

SIX / Strategic Marginality / 115

SEVEN / Peripheral Assent / 139

EIGHT / The Cuban Cure / 161

*Notes / 171*
*Index / 195*

ACKNOWLEDGMENTS

This is a book about the making of a particular form of science on the margins of the global economy. It raises a series of questions about the nature of innovation—questions which traverse traditional disciplinary bounds of anthropology, the history of science, and my own discipline of geography—and how such innovation is shaped in relation to broader political and moral economies.

The research for this book was originally carried out in 2002, when I spent a good deal of time in both Havana and Miami. In Havana I made formal visits to some of Cuba's most important biotechnology centers and spent time chatting informally with scientists and practitioners working at various levels in different parts of Cuba's public health and medical research infrastructure. These contacts provided the bulk of the data used in this book, though I updated some of the material in 2006. While I do seek to draw some more general points from the Cuban experience, I do not therefore claim to offer an account of Cuba's current biotechnology exploits. This book is about the nature and substance of a particular approach to science, written in dialogue with a series of questions of long-standing interest to historians and philosophers of science, anthropologists and geographers. It is not evaluative in the commonsense understanding of that term.

However, the questions raised by Cuba's biotechnology experience have become more, not less, relevant. The financial crisis in the West since late 2008 has boosted interest in how all societies might continue to strive for the "cutting edge" in various fields, but without the same level of investment they might once have counted on. At the same time, within global public health, amid the growing fragmentation of global health delivery into multiple and competing public and private interests, calls for a return to something more closely approximating the Alma Ata vision of comprehensive public health are being made once more with increased vigor. And within the sphere of biomedical research too there is a growing interest in finding ways of making our capacity to enhance and to secure human life avilable to a much greater number of people. All these are questions that the Cubans have a unique and interesting approach to, and, whatever one's views of the political system in Cuba, the sheer scale of global health need today is such that their efforts deserve to be understood. This book is but one attempt to do that.

In writing it, I have incurred a significant debt to many people. First of all, this book is a product of the kindness, generosity, and, in some cases, bravery of a number of individuals in Cuba. Following anthropological convention, all names in this book have been changed, save for those more senior individuals who spoke to me on the record and without whose name my account would not hold together. I am indebted to all alike. When I first proposed the idea of studying the history of Cuban biotechnology my requests were met with polite reservation in Cuba. The initial trust shown me by a few scientists was thankfully passed on, however, and so it is to them that I owe the greatest thanks. They and many others, both in Cuba and in the Cuban-American community, may not agree with some of my arguments here. But I hope they will acknowledge that all are written in the spirit of objectivity. I must also thank a number of public health specialists in Cuba: Gregorio Delgado, Miguel Márquez, and Francisco Rojas Ochoa, in particular.

There were many scientists, academics, administrators, and others who responded to my requests for interviews and shared their valuable time with me, both in Cuba and elsewhere. I am particularly grateful to Daisy Henriques and the staff of the Centro de Salud y Bienestar Humano at the University of Havana for providing me with an institutional base from which to launch these various requests for interviews. Likewise, a good many people at the Center for Genetic Engineering and Biotechnology, the Center for Molecular Immunology, and the Finlay Institute in Havana were extremely helpful, in particular Pedro López-Saura, Gerardo Guillén, Arlene Rodríguez, Rolando Pérez, and Gustavo Sierra. Julie Feinsilver, whose own landmark work on Cuban biomedicine was a major influence, offered crucial early advice on the project. The staffs at the library of the Ministry of Science, Technology, and the Environment, at the José Martí National Library, and at the Hemeroteca in Havana were helpful in providing me assistance despite the difficulties of their situations. My thanks are also due to those who helped at the Cuban Heritage Center in Miami, Florida, especially Maria Estorino. In Miami I am grateful to Stuart Corbridge, who not only offered a place to stay but who was, and continues to be, a challenging interlocutor on all things geographical. There are many others who, for various reasons, I have not named. And there are those whose input, in the form of interviews or advice, I gratefully acknowledge but who I have not been able to reconnect with: David Allan, Germán Rogés, and John Meers.

The majority of the research for this book was undertaken with the assistance of an Economic and Social Research Council (ESRC) PhD studentship award, augmented with generous support from the Smuts Memorial fund and my parents' own coffers. I would like to acknowledge also the intellectual support of Jesus College, Cambridge, and the staff and students at the Department of Geography of the University of Cambridge. I have benefited from discussions with many people there: Andy Tucker, Denisa Kostovicova, Rich Powell, Hannah Weston, Nick Megoran, Nic Higgins, David Lambert and Steven Legg, Olivia Bina, Rory Gallagher, Kaveri Gill, and professors Mia Gray, Phil Howell, and Richard Smith. Jim Duncan and Sarah Radcliffe at Cambridge offered supportive supervisory input, while the Cuba Research Forum under Tony Kapcia, a vital link to research communities in Cuba, helped me begin to excise my initial ignorance of all things Cuban. The work for this book began at Cambridge, but it was completed at Queen Mary, University of London, and my colleagues there have given tremendous encouragement. Most of all, however, I must thank my PhD supervisor and good friend, Gerry Kearns, for his ongoing intellectual guidance and for providing a model of scholarship one can but hope to emulate.

Parts of this book have appeared before. Chapter 4, "Science City," has appeared in a different form in the journal *Environment and Planning A* (where I make a rather different argument about the economic geography of non-Western science), while some of the arguments about development in Cuba have appeared in the journal *Geoforum*. I have given most of the points in this book an airing at numerous talks and conferences, and I acknowledge the advice and pertinent insights of many who attended those talks.

For their helpful comments on draft chapters, I thank David Livingstone, Bronwyn Parry, Felix Driver, Miles Ogborn, and Michael Briant in particular. A number of scientists in the United Kingdom and the United States also gave of their time, insight, or advice much later in the project, particularly a varied group of immunotherapy experts including Paul Harari, John Gribben, Gustav Gaudernack, Sandra Diebold, and Herman Waldmann. Similarly, at the Wolfson Institute in London, I am grateful to Salvador Moncada and to Miriam Palacios-Callender. At the University of Chicago Press, I am grateful for the wonderful and patient input of my editor, Christie Henry, and Susan Olin's careful copyediting. Though now relocated outside of Cuba, my thanks to the Vedado crowd of Steve, Yañia, and Marisol. Though now also relocated from where we first met in Havana, I am grateful to Jens Plahte at the Norwegian Institute of Technology for his engagement and informative discussions over the years. Thanks above all to my family for their constant support, and to Katerini, who has put up with my inability to put this down for longer than she might care to remember.

INTRODUCTION

In July of 1986 a number of scientific luminaries, journalists, and the directors of both the World Health Organization (WHO) and the Pan-American Health Organization (PAHO) gathered on the outskirts of Havana. They had assembled for the opening of what was then one of the largest research laboratories in the world: the Center for Genetic Engineering and Biotechnology (CIGB). The opening of Cuba's Centro Genético, as it would come to be known by many Cubans, was intended to stand as a pivotal moment in the history of Cuban science. Since the late 1970s the unfolding biotechnology revolution in the United States and Europe had shown the huge potential of this field: a genetically engineered human vaccine was already available, and the first field tests of genetically engineered plants had recently taken place. Indeed, it was the promise of just such developments that had led Fidel Castro to decide, four years earlier, that Cuba too needed to develop a biotechnology industry.

The opening of the CIGB was to mark Cuba's arrival as a biotechnology player on the international scene. Television cameras were there to cover the president's speech. A reporter for the popular magazine *Bohemia* introduced the new center to its Cuban readership, describing "the majestic and sober architecture" of the buildings, which "truly caught the attention."[1] The journalist evoked a sense of magnitude and awe, of mission, and of promise. *Granma*, the official state newspaper, printed the comments of many of the visiting dignitaries to the opening event, including the representative of UNIDO—the United Nations Industrial Development Organization—who concurred in "marveling" at the center.[2]

As the presence of these other figures suggested, the opening of the CIGB was also intended to be a pivotal moment in the development of high technology science for other developing countries. Hence, it was not just Castro who had set considerable store by the intended achievements of the new center. As the biotechnology representative of the United Nations, Rodolfo Quintero Ramírez, would later admiringly declare, this institution "could be situated in any place in the world."[3] But the realization of this science-led development was a vision that Castro had advocated more vocally than most, and in fact the CIGB had a very specific Cuban dimension. Castro had declared in 1961 that "the future of our country has to be a future of men of science, it has to be a future of men of thinking precisely because this is what we are now sowing."[4] By 1986, that phrase had become a leitmotif of the revolution. For Castro, therefore, the opening of the CIGB had another purpose: it was intended to demonstrate that the political revolution in Cuba was about to converge with the scientific revolution in biotechnology.

## Experimental Modernity

In 1940 Cuban anthropologist Fernando Ortiz wrote a seminal book called *Cuban Counterpoint*—a study of how the island's switch from a tobacco economy to a sugar economy precipitated a series of cultural convulsions throughout society at large.[5] The subsequent attempt in Cuba to realize Castro's hopes for biotechnology was pitched in a not dissimilar light. When set against developments elsewhere, the emergence of biotechnology on this small Caribbean island over the last twenty-five or so years is certainly more than a little remarkable. For reasons of necessity (such as limited finance) and contingency (such as the desire to apply the results for political ends), Cuban scientists have developed an alternative approach to biotechnology that responded to local needs and that was nurtured within a different space of ethics and possibility. But it would be foolish to say they have remained entirely immune to the waves of hype that capital-driven biotechnology in the West has tended to generate. The events recounted in this book, which sets the history of some of Cuba's biotechnology endeavors alongside an analysis of the nature and broader implications of that work, are very much about the points of intersection between the globalization of a Western scientific rationality, on the one hand, and a distinctively local "experimental" milieu, on the other.

Biotechnology is a very modern science. Offering the potential to invent and to then patent life itself, biotechnology has also always been a heavily capitalized science because what is patented can be marketed. It has borne, at the same time, the hopes of many for miracle cures and the anxieties of others about the consequences of our manipulation of living matter. But, beneath all the hype, biotechnology has also elicited novel ways of thinking about and relating to biological and social life. The first patent for a genetically altered animal was awarded in 1988; a draft version of the map of the human genome had been produced by 2002. The effects of this are substantive: the various technologies, techniques, and institutional forms that make up what we understand by "biotechnology" have been made possible by, and have in turn helped fashion, new forms of social organization, scientific and political practice, institutional frameworks, and legal and financial networks. Biotechnology has become, in short, a quintessentially modern assemblage.

Some have taken the implications of this quite far. Thus society now stands, so they claim, on the brink of a new "posthuman future."[6] Such claims as to the novelty and historical importance of biotechnology tend to overlook a prior question, however: namely, how are our competing knowledges and ways of doing biotechnology deemed to "count"? It may seem naive to ask who gets to decide which research questions, which methods, which styles of research, which ethical frameworks are to be valued over others. But modern assemblages like biotechnology do not simply emerge; they are put into place by particular interests working in a given context, be those interests scientific, regulatory, or of the public itself.

The standard narrative of biotechnology, for example, understands it to be the product of a particular place and cultural milieu: America in the 1970s, where a combination of university research, private investors, and political authorities capable of ensuring an amenable set of regulations all fell into place. The well-documented 1980 US Supreme Court decision, in the case of *Diamond v. Chakrabarty*, for example, which upheld genetic engineer Ananda Chakrabarty's patenting of a bacterium capable of breaking down crude oil, opened the door to proprietary ownership of living matter and a gathering slew of patents which today sees new biological entities "invented," registered, and owned every week. If such elements constitute what is taken for the basic architecture of today's biotechnological and pharmaceutical industries, then this is an architecture held together by the sort of "American exceptionalism" first identified by Alexis de Tocqueville: a cultural glue encompassing a strong association of newness with improvement, an imperial relationship to nature, and substantial capital investment.

The form in which contemporary biotechnology has developed has thus been a very specific, marketized form, sustained by a growing trade in biological entities and their informational proxies—be it emergent forms of digital classification or simply the electronic record of shares traded daily on the NASDAQ biotech index. In 2005, for example, the global biotech industry raised a record $19.7 billion in capital.[7] The science of biotechnology is acutely accountable to these capital interests—to the venture capitalists, the university regents, and the experts who navigate the difficult path from the science that is viable in the lab to the product that is viable in the marketplace. As these individuals and their products and ideas crisscross the world, squaring regulatory demands with shareholder commitments, and translating laboratory science into corporate action, they have marked out a distinctively capitalist web of intellectual property rights, markets, regulatory norms, and international protocol around biotechnology science.[8]

Events such as the annual Biotechnology Industry Organization (BIO) meetings, for example, are dominated by strategic information on the latest partnering opportunities for European and American companies, and the principal lexicon is that of business and management. One gets relatively little sense of the sorts of developments that are taking place outside of these "central" locations, however: in India, China, and Brazil, for example—where fully three-fifths of the world's industrialized population live. And yet, China is currently developing the largest plant-biotechnology program outside the United States, while India and Brazil also both have impressive and growing biotech capabilities.[9]

While there is some recognition of this at the BIO events, there is also a slowness to respond to it. Countries as diverse as Morocco, Vietnam, Argentina, Kenya, Costa Rica, and Cuba all have active domestic biotechnology capacities, yet there is often a lack of interest in the work going on in these countries. Headline news at BIO 2006, for example, was that biotech in Asia now accounted for 46 percent of total industry growth. Despite the impressive figure, very little was said about the sort of work or achievements that comprised that 46 percent growth. There is a double problem here, because one of the reasons these countries often may be far from the cutting edge of research is that the same legal, economic, political, and cultural forms of sanction that structure biotechnology as a global apparatus also go into regulating and controlling it, with those at the center of this system being those who have most say in its rules. Those who started late, or who do things differently, struggle to make their mark.

The specific question I set out to ask in this book then is, if biotechnology does indeed articulate many of the features we associate with a distinctively modern Western capitalism, are those features *necessary* (that is, do they constitute it in some way), or do they simply lead to just one of numerous possible formations, numerous possible "ways of doing" held in place by interested parties? What is there to learn from some of the alternative approaches to biotechnology that exist in the world? In short, what of the local in these modern and universalizing formulations?

## Epistemic Articulations

It was just this question of the role of the local that surfaced most strongly as I spoke to delegates at a meeting at the same Center for Genetic Engineering and Biotechnology in Havana in November 2001. The meeting reflected on fifteen years of prior work on interferon in Cuba and around the world. The director of the CIGB, Luis Herrera, opened the event with a keynote address: "This is a proud moment, a marvelous moment in the history of [the] development of this field," he said. Herrera spoke of the need to advertise Cuban work more broadly in the future. For the benefit of the international guests, he also recalled the past: of how Cuban biotechnology began, somewhat modestly, in a converted villa staffed by just a handful of scientists, and of how, just a few years later, those scientists had carried out innovative work on the therapeutic uses of interferon. They had toiled night and day to produce a recombinant hepatitis B vaccine and had successfully completed a scientific mission that would bear perhaps the ripest fruit of all—the VA-MENGOC-BC—to date the world's only effective vaccine against this disease.

Situated on the periphery of dominant Western imaginations of biotechnology's development, the Cuban attempt to develop a form of noncapitalist, bottom-up biotechnology not only represents a novel variation on Western biotechnology practice but one that runs against the grain of biotechnology's presumed universalizing formulations. That much many of the delegates at the event in 2001 were convinced of. But where, precisely, did the difference lie? Here, responses were more mixed. One clue as to why may be found in the famous address "Science as a Vocation," delivered in 1918 by the German sociologist Max Weber. In it, Weber sought to develop an understanding of what he termed "conditions of science."[10] For Weber, a country such as the United States—and there were a good few Americans in Havana to judge of this—represented a classic democratic model of science. Compared to Weber's native Germany, say, a scientist in America was not so likely to require a private personal income to support a burgeoning scientific career. But the price that the individual paid for this was expendability and an overburdening of teaching duties. The German system, on the other hand, was more hierarchical and less considerate of individual liberties, but for that, Weber saw, it fostered a greater depth of clarity and the patient exploration of one's own intellectual depths.

Mingling with the scientists at the meeting, it might have been tempting in some ways to recast Cuba and America in these contrastive roles. Certainly Cuba represents a strongly polarized example of scientific ideology and "conditions of science" relative to America. The culture of research ascendant in the United States continues to emphasize scientists' roles as individuals, their careers accountable ultimately to themselves and dependent on private sources of funding. In Cuba, however, scientists are supported by the state and are therefore expected to contribute to whatever projects are determined to be in the greater good. The most important differences, however, are with respect to that elusive but powerful notion of "clarity." By clarity, Weber famously referred to the way in which the scientific subjectivity sought to give "an account of the ultimate meaning of [one's] own actions." Accordingly, the role of science in society is to provide clarification concerning these "ultimate" problems, Weber suggests, and this proves to indeed be rather differently thought in Cuba as compared to the West.

As he spoke to the gathered scientists at the opening of the CIGB in 1986, Fidel Castro sought to impart his own meaning to precisely this task. He similarly focused on the scientist's role in society (on the notions of "duty" and "responsibility") and similarly (though not, it seems, intentionally) placed an emphasis on Weber's notion of "clarity." Castro, however, had something rather more specific in mind. Emphasizing the nature of scientific conduct that he saw within Cuba's scientific community, Castro proudly declared his confidence in the 1,200 or so scientists gathered around him. They would work, he suggested, for the benefit of science and the nation. Cuban biotechnology science was to be a *collective* endeavor, then, conceived as contributing to socialist society more generally. But the scientific "truths" it was to establish, the discoveries it was to make, were to be formed in the crucible of local, immediate, and distinctively *national* needs. He even went so far as to refer to the scientists as "our priests of science." Priests in Cuba are more usually associated with the island's Santería religions, and their work is more closely attuned to notions of magic than to the doctrines of Western Christianity. It was a highly complex as well as rhetorical claim, then, but the parallel is indicative of the excitement of the day. In many ways these scientists were similarly being called on to work their magic and, accordingly, Castro spoke of miracles: "We have preserved the DNA of the famous cow, Ubre Blanca [White Udder]" he declared. Ubre Blanca now graces the foyer of the Institute for Animal Health in Havana—a sizable and prolific Holstein, she was capable of producing over seventy liters of milk a day. "Perhaps it will be possible one day to reproduce this cow," Castro mused.[11]

Such words set great store by the promise of biotechnology, but they actually underplayed what was in fact the more pertinent issue raised by that day's proceedings: the challenges of developing a paradigmatically *capitalist* form of science in a developing socialist country. This was of course partly the challenge of finding the vast sums of money required even for a stripped-down biotechnology project. But for the same reasons that biotechnology in America can in many ways be pegged back to de Tocqueville's cultural diagnosis, it was equally, if not more so in Cuba, an epistemological challenge. Despite problems that will become apparent, the Cubans' experiences in developing an alternative space of scientific endeavor and, in particular, the forms of ethical practice that underpin it, illustrate what it means to come up against the cultural biases and constructions that lie hidden in this most high-tech and presumed "universal" of sciences. Their experience raises two broader theoretical questions that I seek to examine in this book. First, to what extent are forms of scientific reason and rationality geographically bound? And, second, what happens when different forms of scientific reason collide?

## Placing Science

In an attempt to get at the nature of scientific advances, the philosopher of science Thomas Kuhn argued that theory change in science is not a matter of accumulation of knowledge—of new theories bettering the old—but of changing intellectual circumstances and possibilities.[12] It is a model whose principal vector is time. If we question the articulation of such circumstances and possibilities across space as well, however, we begin also to raise questions about the limits to any given epistemic community (such as we have begun to consider above) and thus about the effects of interactions between them. We begin to raise questions, that is, about the relationship between knowledge making and place making in science, and its implications for the political economy of biotechnology.[13]

The philosopher Henri Lefebvre reminds us that place making is not simply the product of clearly articulated intentions. It arises as a contingent element of social life itself. The places we create are the product of the imperfect realization of our attempts to shape those places to our needs. The same is true of attempts to produce knowledge in the laboratory or in the field: "The locales in which scientific knowledge is produced are not seen as passive backdrops but as vital links in the chain of production, validation and dissemination," state the editors of a special issue on the "geography of truth."[14] These places thus also shape, in important ways, our conduct within them.

Given the geographical variations in the practices whereby drugs are moved from laboratory to patient, some of these spatial insights—which I want to elaborate later on—are particularly pertinent to the study of global biotechnology. The practice of pharmaceutical companies providing doctors with their own drugs is routine in the United States, for example, but illegal in many other places, including in Europe where it is nonetheless raised by the pharmaceutical lobby every year. This presents basic irregularities in regulatory space—the way that normative principles are locked into regions and territories through the always limited reach of different legal systems—that scientific research and the marketing of that research in the form of products must take into account. The same is also apparent in the geographical construction of notions of "need" that help to determine who can produce what sorts of drugs in the first place: in 1998 and 1999 the US government prevented South Africa from using the national emergency clause in the TRIPS (Trade Related Aspects of Intellectual Property) agreement to import AIDS drugs, but then invoked the same clause itself to bolster national competitiveness in microchip production. Here, what was a matter of life and death in one place was seen as a question of competitiveness in another. To tell stories of science that leave out the "where" is, I would therefore suggest, to tell but part of the story.

To be sure, there has long been an explicit consideration of spatiality within many social studies of science, and it is not the task of this book to try and situate this rich body of work within some overarching framework of the "geography of science." Attempts to understand the production of knowledge have, for example, consistently invoked certain clearly geographical dynamics (such as notions of "disembedding" deployed by geographically minded economists) or they have examined scientific developments taking place across implied geographical terrain (through what sociologist of science Bruno Latour calls "chains of translation").[15] Important insights accrue from this: scientific knowledge is modified as it travels; forms of standardization can be hard to reproduce. As David Livingstone has shown, even the reception of scientific knowledge is geographically variable: Darwinism supported racial ideology in colonial New Zealand while undermining it in the American South.[16] My approach to the story of biotechnology in Cuba engages with these sorts of questions. But it seeks primarily to draw attention to the geographical elements that, though they may be integral to any particular historical narrative of science, are often hidden within them. It aims for a sociology of knowledge that seeks to make such inherent geographical variability its starting point rather than its backdrop.

Such a spatialized epistemology ought to proceed, it seems to me, in at least four registers. First, it ought to elucidate not only histories of scientific endeavor but the geographical conditions of existence for such endeavor also; second, it ought to assess how different spaces of knowledge impact each other and to probe the formation of scientific relations within, across, and between them; third, it ought to elucidate the deployment of specifically geographical reasoning within these emergent epistemological cultures; and, fourth, it ought to take into account how such knowledges are differentially articulated. There are both empirical and conceptual implications of such an approach that offer clues to the nature of biotechnology research in Cuba. At the empirical end of the spectrum, for example, an interest in the "life sciences" ought to pay attention to the ways that both "life" and "science" mean different things in different places. At the conceptual end, by extension, these very articulations of "life" and "science" designate a broader assemblage of elements held together by particular strategic logics. The clinical trial as an institutional form of warrant making, for example, is dependent on ethical norms that are themselves determined by particular, situated interests, be it the shareholders who have invested in the drug on trial or the scientists committed to seeing the fruits of their labor.

To understand the full implications of the Cuban story of biotechnology science, therefore, means first gaining a sense of its particularity in this deeper sense; it must be located with respect to broader developments, and the underlying forms of rationality which structure that arrangement must be revealed. In short, it means asking exactly *how* is biotechnology in Cuba different? Answering this question offers insight into both the nature and organization of power in the global biotechnology industry and the problems and possibilities faced in constructing alternative scientific rationalities. It also helps elucidate the geographical unevenness with which the power to make knowledge and things is distributed. Such uneven geographies are important because they are central to the production and maintenance of the current global biotechnology industry and the broader neoliberal pharmaceutical regime of which it is a constitutive element.

## Reason and Resistance in Global Science

Elaborating such a geography is central to the entire story I tell in this book—whether the initial chapters, principally about the "making" of biotechnology in Cuba, or the later ones, where I broach some of the implications of these observations about the fashioning of science in place and the geographical constitution of knowledge. Despite still popular characterizations of science as a solid march to universal truth, scientific knowledge is the product of struggle—a fact that scientists may be more aware of than the sociologists who study them sometimes give them credit for.

Such struggle becomes particularly acute when the grounds on which scientific knowledge is produced are themselves open to moral, political, or other vocabularies of suspicion. The Cubans confronted precisely these problems in the 1990s when, having successfully put together an alternative scientific milieu in the 1980s, they sought to reconnect this particular space of experiment with a globalizing and avowedly postsocialist world—in short, when they sought to make their knowledge count elsewhere. And as they would eventually find at the hands of the US State Department, the question of whose knowledge counts is too often answered by means of a simple shorthand—"where does this knowledge come from?"

The relationship between reason and the broader forms of rationality that structure it is thus an inherently geographical one. Reason, in the sense I use it in this book, is what is daily deployed in the conduct of scientific as well as other personal and professional endeavor. It is "the power of the mind to think, understand and form judgements by a process of logic." Rationality, on the other hand, connotes the norms on which such reason is based. It is not reason of itself, but rather it explains how action is taken in relation to reason and illuminates what is determined as being reasonable in any given situation. It is the constitutive context in which reason is deployed. In this light, the Cuban case raises some intriguing questions about the operation of science as a social apparatus: a geographically uneven assemblage of preferences and probabilities as much as one of competence and technique. It also prompts us to consider how resistance to the dominant rationalities that hold particular forms of reason in place might open up.

In order to explore this, chapters 1–4 take a holistic look at the "making" of Cuban biotechnology in the 1980s and the sorts of practices that were involved: discursive, political, architectural, legal, and so on. The intention is that an impression of what might be termed, cautiously, the heyday of Cuban biotechnology emerges from these chapters. By paying attention to the made qualities of this science, I try to trace out the ways that a particular form of scientific rationality was articulated in the daily work of Cuban biotechnology scientists and to assess the extent to which this contributed a distinctive element to the practice of biotechnology in Cuba.

Chapters 5–7 then examine various points of intersection between this particular "Cuban" system of thought and the global norms and regulatory systems particular to the West (outside of which the Cuban system had been formulated, and against which it would now be *re*formed). This part of the story really turns on two substantive issues: how we establish warrant for the production of truth and how we negotiate different forms of accountability. The first of these concerns the cultural values that underpin decisions as to which forms of knowledge are allowed to count. The second concerns the cultural values that determine which of those determinations count.

The story of what has happened in Cuba over the last two and a half decades, set against the development of biotechnology as a distinctively Western global apparatus, is thus ultimately a story about the obstacles facing poorer countries in overcoming their pharmaceutical dependencies. It is about the practice of global science, the nature of regulation, and the geography of reason and resistance in global biotechnology.

ONE

# A Biotechnology Story

It is late November 2001 and I am talking to Pedro López-Saura, one of Cuba's foremost experts on interferon. López-Saura is a vice director at the island's flagship biotechnology institution, the Center for Genetic Engineering and Biotechnology (CIGB). We are seated at an international conference reporting recent work on interferon and cytokines. Interferon itself has made its own contribution to the larger story of biotechnology science, and the Cubans have worked on interferon since the very beginning of their interest in biotechnology, he tells me. Indeed, it is almost impossible to find an account of Cuba's genuinely remarkable biotechnology project that does not on some level fall back on the agency accorded this innocuous protein, which is naturally produced by animals in response to infection. But for all that this is a popular story, it is a narrative that obscures as much as it reveals.

## Fast Science

Interferon was first discovered over fifty years ago, in 1957, by Jean Lindenmann and Aleck Isaacs, who had been studying how cells are protected from infection by one virus as a result of prior infection by another.[1] Subsequent work revealed the role of interferon in this process; it worked as an immune system regulator. It was not until the late 1960s that interferon first caught widespread scientific and public attention, however, when an American researcher in Paris, Ion Gresser, discovered that interferon also stimulated the production of tumor-killing lymphocytes (white blood cells) in mice. And it would be a further decade before interferon could be obtained in sufficient quantities to support any greater experimental, let alone clinical, practice. As Lindenmann himself later noted: "Beginning as a rather esoteric laboratory phenomenon . . . [interferon] passed through a sort of cottage industry to end up part of the metropolis of modern biotechnology."[2] By the late 1970s, when the Cubans first took an interest in interferon, it was still curiously poised, therefore, as a drug whose inherent properties many thought might make it a magic bullet—perhaps even *the* drug to beat cancer—yet one which was also extremely hard to produce.[3] Like everyone else working on interferon at the time, the Cubans would have to figure out not only how to produce interferon in theory; they would also have to solve the problem of producing it in sufficient quantities to actually do anything with it.

One of those heavily involved with interferon research at the time was American oncologist Randolph Lee Clark—president of the M. D. Anderson Hospital in Houston, Texas: America's first cancer hospital. Clark had set up the Anderson Hospital back in 1946 when, as a young physician emerging from the war and fascinated by the global scale of modern medicine, he developed a lifelong international outlook. By his later years he had become a sort of scientific statesman. He advised President Nixon on cancer research and he traveled often, regularly touring medical facilities abroad. When he heard in late 1980 that a tour of Cuban health facilities was being proposed by US Congressman Mickey Leland, he took little persuading to come along. "Healthcare has no dialectics," a colleague of Clark's said when inviting him to come along. "We, Cubans and Texans alike, have common goals in caring for our people."[4]

The idea behind the visit was to initiate a series of scientific exchanges between the two countries, outside the political stalemate of the time; an impetus that dovetailed perfectly with Clark's enthusiasm for science in developing countries. "This is a real 'going Jessie,'" he confided to his diary on the last day of the visit, "hinged on one man, as far as I can tell. [Y]ou're going to hear a lot from him—he's going to be a sounding board but he's going to talk the language they need to hear in the third countries." He was referring to Fidel Castro, and during the trip the two men met face-to-face. Clark had dedicated his career to searching for a cure for cancer. That he had begun his work in a converted stable on a donated family estate appealed to Castro's sense of the dedicated professional. Castro impressed Clark too. Not least, the two of them had the diplomatic instinct for the globally unspecific. "A bit vague?" one of Clark's students is supposed to have remarked. "Vague is much too precise a word."[5]

By the time Clark left Cuba, he had rather more certainly left Castro with the sense that interferon was *the* drug to be working on.[6] He had also offered to host a guest researcher from Cuba at his hospital, to which Castro responded by asking if Clark would be kind enough to take two. While the arrangements were finalized by Clark's office in Houston, the Cuban interests section in Washington, and Castro's office in Havana, Clark forwarded the latest in scientific knowledge on interferon to Cuba, where it was channeled to some of the country's most promising scientists.

López-Saura was one of these scientists, but it was two other researchers, Manuel Limonta and Victoria Ramírez (a hematologist and gastroenterologist respectively), who were chosen for the trip to Houston. The two Cubans worked solidly for over a week there. Limonta's concluding report points to the exhausting training they had received—peppered with the occasional perk, including a flight over Houston by helicopter. Evidently enthused by his experience, Limonta penned an informal note at the end of his report that is almost as telling as his accounts of their growing experimental proficiency with the drug: "The advise [sic]," he commented, was of particular use "not only because of the great experience and knowledge of the different fields of medicine he [Clark] has achieved in so many years of intensive work, but also because of the great sensibility and deep wish to help humanity." In other words, what the Cubans had just been taught about interferon was not just valuable as scientific knowledge; it was valuable for its social content too. Indeed, Clark had also informed the Cubans that the best place for advanced work on interferon was in the Helsinki laboratory of Kari Cantell, where they were to head next.

Kari Cantell had been the first to isolate interferon from *human* leukocytes back in the seventies. From his laboratories at the State Serum Institute in Helsinki, Cantell produced the raw material of natural leukocyte interferon for use by researchers around the world. Once Limonta and Ramírez returned to Cuba it was, accordingly, to Helsinki that another working group of *cuadros*, as they are known in Cuba, was organized to travel.[7] Cantell's method was, after all, widely accepted as the industry standard. It was this second group that Pedro López-Saura was invited to join: "The group was made up of six people," he told me.

> We were all doctors, but three of us were also biochemists. I had been working for a while on a project to improve the quality of clinical studies and also for a while on yeasts when this issue of going to work on interferon arose, in early 1981. We spent one week in Cantell's laboratory, observing the techniques involved in the process. Afterwards we took another week [as] it was necessary to make inventories and to purchase some of the equipment and materials needed.[8]

Cantell was famously generous with his purifications. One of the few able to produce significant quantities of high-quality interferon, Cantell also viewed it as a basic resource to be shared. Famously, he never took out a patent on his procedure—which was why, in the hype that followed initial results suggesting interferon's possible antitumor properties, the American Cancer Society was able to initially order $1 million of interferon from Cantell, only to purchase but a third of it from him and the rest at twice the price from an American company set up using Cantell's own method and owned by the president of the self-same American Cancer Society.[9]

With Cantell, the Cubans had again found a form of scientific solidarity that cut across the political constraints to their desired line of work. And with them Cantell willingly shared not just the product but the means for the Cubans to make it themselves. In consultation with Cantell, a former luxury house in Havana was fitted out as a laboratory with all of the equipment that Limonta and his group would need to produce interferon. The six scientists began working in the laboratory, known simply as House No. 149, the day after their return to Havana. There, for several weeks, Limonta, López-Saura, and the others set about retracing Cantell's method according to his own instructions. As an added incentive perhaps, Castro, who had taken a personal interest in the project, would often stop by in the afternoons. He personally saw to it that the scientists were supplied with whatever they needed. The costs of such a small-scale and dedicated approach were evident, of course: an extremely narrow focus based, for the time being, on just one drug. But the compensating advantage was speed. While it had taken some groups up to a year to perfect Cantell's art of purifying interferon, it took the Cubans just a few weeks. Limonta's letter to Clark, informing him of their progress, put it succinctly: "On April 12 the group arrived in Cuba after twelve days training. On April 18 a place was adapted to start the laboratory work. By the middle of May the first purification of interferon was done."

Just a few days after the Cubans made their first breakthrough with interferon, the island was struck by a virulent strand of hemorrhagic dengue fever, which at its height saw 11,000 new cases diagnosed each day and over 116,000 people hospitalized in total. It was the first of this particular strand of dengue, for which mosquitoes act as the transmission vector, to have appeared in the hemisphere.

The outbreak prompted suspicious words by the Cuban government directed at the United States, while the country's health system was mobilized in response. But even Cuba's extensive health-care institutions were unable to cope. Some form of active medical response was needed, and some looked to the recent work on interferon. A then director of one of Havana's principal schools of surgery, who I met on the veranda of his institute in the Vedado district of Havana, explained what happened next.

> "I wrote a letter to the Ministry of Public Health stating my interest in the idea of using interferon in children with acute hepatitis B: we were a long way from producing a vaccine for it then . . . but I had started to apply interferon with Manuel Limonta on sporadic cases of hepatitis, visiting patients between 11 and 1 in the morning because we were in the lab all day actually producing interferon. We had to be extremely careful—there was no publication with which we could work by, so we had to be extremely careful given the possible side effects."[10]

Tests using interferon were given the go-ahead and initially undertaken on three hundred people, over half of whom were children. A tense few days passed before early results showed the Cuban-produced interferon to have a marked immunoregulatory effect.[11] Shots were immediately administered as a prophylactic to the most vulnerable, and within a few months the epidemic had largely died down. It was, as the government has since represented it, an "epic" battle between society and nature which soon also saw the mosquito stamped out of thirteen of the country's fourteen provinces in an environmental health program overseen by the authorities.[12] Interferon may therefore not have been a central part of overcoming the illness (which quite arguably owed more to rapid diagnosis, the early hospitalization and rehydration treatment provided, and the mass involvement of 10,000 volunteers). Nevertheless political awareness of these modern medicines had been raised, along with a sense of how they might play an important role in addressing the nation's problems.[13]

At this point it pays to stand back from this narrative for a moment and to reflect on some of the developments afoot as this small troop of scientists sought to develop interferon into a lifesaving formula under scrutiny of the state. Such was the national scrutiny, in fact, that their work was also somewhat removed from most forms of international oversight. The Cubans' fast-track approach to developing new drugs required breaking down the boundaries between experiment and application, a controversial approach justified on the basis that any potential risks were outweighed by the likely social benefit. Such an unusual degree of overlap between the clinical development of the drug and its application was possible because the scientists undertaking the research were also the doctors administering it to patients. The spaces of the laboratory were thus in a certain sense being folded into society itself. Such an approach resonates today with more recent arguments about the need to rapidly develop drugs that have a clear public health application (with respect to debates focused in particular on the clinical application of HIV drugs prior to the full demonstration of clinical efficacy and safety through completion of randomized clinical trials). But in 1981 the Cubans were early starters with such an approach. Moreover, the linking of clinical application with research in the context of the "war footing" of the dengue epidemic did not remain unique; it quickly became the norm.

The emergence of hemorrhagic dengue thus provided the impetus for a change of direction both in the clinical development of interferon in Cuba relative to other countries and in the experimental ethos of Cuban work on biotechnology.[14] The application of interferon was an uncertain process, but the state's need to respond effectively to the public health crisis meant that standard regulatory processes of proving efficacy and safety were partially suspended in order to accord primacy to the demonstration of population-level change and, more important, to enable a rapid response.

In short, the scientists involved in this particular event were given greater freedom in order to ensure that whatever the means, the ends of overcoming the crisis were met. A clear difference between this early foundational biotechnology work in Cuba and that taking place elsewhere was thus already emerging: biotechnology in Cuba was driven by public health demand (which, as one informed observer put it, "is highly unusual: it is normally a case of *industry* into biotech") rather than a push for innovation and profitable research from within science.[15] In place of an economic imperative, therefore, were the social demands that the Cuban scientists apply the products of their research as soon as possible. What Clark and Cantell had proffered as a maxim, the Cubans had thus developed as a methodology.

This was not lost on Clark. In October of 1981 he met briefly again with the Cubans as they passed through Texas:

> [Limonta] was on his way to San Francisco to meet at the interferon conference there this weekend and I arranged for them to see Hersh and Gutterman thinking that they should know of this opportunity to work with Cubans who would have much less regular constraints in their advancement of clinical trials, and would have the whole of the Cuban population to call upon. Apparently this is backed directly by El Presidente and he sent his regards today.

Having committed his thoughts to Dictaphone, Clark then sat down to respond personally to Castro. "Dear Mr President," he wrote. "It was indeed a pleasure to receive your emissaries, Drs. Manuel Limonta and Luis Herrera, this week and to learn about your remarkable progress in establishing a laboratory for the production and use of interferon." The business dealt with, he moved on to the personal element. "May I also take this opportunity to thank you for the magnificent wooden box filled with your special Havana cigars, the record album and the bottle of Cuban rum . . ." Rarely had scientific diplomacy been so simple and yet so acute.

## Slow Paradigms

"It is tempting to begin by describing what was known," writes François Delaporte at the start of his essay on the birth of tropical medicine. In uncovering the emergence of a particular approach to biotechnology in Cuba, Delaporte's temptation must indeed be resisted. The above narrative recounts the emergence of a rather particular approach to science. Certainly it is true that, through a form of international solidarity, the Cubans were able to obtain the intellectual and material resources they needed to successfully manufacture first-quality purified interferon and, ultimately, recombinant engineered interferon as well. They put this globally disparate suite of work together into a highly focused project that found an immediate application for use in dengue patients in Cuba. But that this story says all that needs to be said about what it is that makes Cuban science different is manifestly not the case. What is most interesting in the above account of the way Cuban work began and developed through an initial specialization on interferon, in fact, is what lies silent within.

Clark's letters may well have been rather matter-of-fact about the Cuban endeavors, but their model of fast science was, as I have intimated, very much against the global trend of the time. It warrants considering that context in a little more detail. In 1975, many of the world's leading scientific authorities met in Asilomar, in Pacific Grove, California, to discuss an unprecedented proposal: a temporary halt to rDNA experiments until the safety issues concerning such research were clear.[16] This was famously an initiative led by the scientists themselves in an effort to address the social and political implications of their work. It was a task that brought their own understandings of that work to the fore. Reflecting on this elsewhere, the anthropologist Paul Rabinow has observed that "a significant omission from the by now classic laboratory studies has also been the representation of science as a practice and a vocation—*by its practitioners*. Such representations must be elicited . . . [and] framed in the light of larger forces at play."[17] It is a pertinent point, to which we might add that such representations as at Asilomar ought not only to be set against their context (framed in the light of larger forces, that is). So too ought we to examine how they are a constitutive *part of* that context.

Asilomar was a classic example of how scientists' self-representations do not just mirror society but help to structure it. Out of the particular self-understanding that Western scientists took to Asilomar was created a narrative of containment that continues to inform public debates over biotechnology today. The interferon story in Cuba, a self-narration that justifies the rapid approach taken to biotechnology there, is another such representation of science. It too forms a part of the world it seeks to describe because it contributes to, indeed becomes, the dominant discourse about biotechnology science in Cuba; it has become, so to speak, paradigmatic.

But it is through such continual enunciation that this story itself has introduced a number of interesting elements into the picture. The anthropologist Kitty Abraham has written of the "unstated anxieties of post-colonial science." For the Cubans working on biotechnology today, there are rather specific things that they might be justified in being anxious about. The speed with which they sought to take Clark up on his offer and to visit those scientists they could were in part a response to the restrictions on Cuban scientists imposed by the United States. But a more specific effect of such *reiteration* is to separate out a discrete story about science (how the technology and understanding required to undertake biotechnology was put together in Cuba) from a messier and more complicated story about politics (that this had to take place under rather specific circumstances resulting from the island's geopolitical positioning). Like Asilomar, the interferon story is something of a slow paradigm, therefore, because it solidifies a certain historical amnesia while articulating a broader, spatialized ontology as to what Cuban biotechnology science is and isn't. From this follow three crucial implications.

First, by claiming it is an exceptional story of scientific triumph in the face of natural adversity, explanations for the Cubans' success with biotechnology that fall back on the interferon story as a shorthand have in fact tended to decontextualize Cuban biotechnology from the very specific conditions in which it occurred. This effects a displacement of agency from the object world to the subjective world of collective will. The implicit claim has been that Cuban scientists were good enough to be able to develop biotechnology despite all the apparent difficulties facing them in this task.

In fact, as we shall see in chapter 2, there are a number of contextual factors, specific to Cuba, which gave the Cuban scientists certain distinct advantages when it came to developing biotechnology.

Second, accounts that focus on the role of interferon, whether as a model for biotechnology in Cuba or as a process of "learning by doing," also serve an official self-truth as to the socialist character of Cuban science (that it was somehow inherently socialist). This self-truth has played a hugely important role in Cuban biotechnology. It tells us, for example, that only a socialist science could have triumphed in the face of adversity in the way described above. But this of course only serves to mask the importance of other sources of value in bringing the social forms that arose from those conditions into a highly novel assemblage. By far the most important of these, as chapter 3 considers, has been nationalism. This effects a geographical boundedness to what is inherently more mobile and stands in contrast to the reality of what the Cubans have achieved.

And third, while these accounts make specific claims for the importance of the political will of the state (*la voluntad política*) as a support for the interferon project, such claims are silent about the effect that such politicization had on the management of that science or on the direction that it subsequently took. This effects an elision of individual subjectivity with the state and in so doing overlooks, as chapter 4 will show, the productive capacity of unintended effects and even resistances that such political will habitually generates.

What the interferon story really reveals, therefore, is the felt need of Cuban scientists to place the emergent biotechnology industry in Cuba within a presumed universal history of scientific advancement. But what that felt need itself expresses is a profoundly local articulation of those global imperatives. So when the CIGB's vice director Carlos Borroto suggested that "the strong Cuban program of biotechnology in the 1980s coincided with the 'boom' in this sector,"[18] he is, on one level, quite right. Cuba invested considerably in its biotechnology industry in the early to mid 1980s, and this was certainly a more general period of rapid development of genomic knowledge and technological capacity. On another level, however, he is quite wrong.

As I shall argue in the chapters which follow, Cuban biotechnology neither boomed nor coincided. Rather, it struggled into being and for some time was entirely removed from the rapid advances in biotechnology taking place elsewhere in the world. In place of a boom something rather more modest and considerably more interesting took place: biotechnology developed there for different reasons and in a different way than had been the case elsewhere, certainly in rich Western countries. This was to have profound effects both on the science itself and on the deployment of that science within the revolutionary machine. But I begin with this story of self-vindication—which is what this interferon story is all about—because the Cubans are quite justified in deploying it. As the philosopher of science Ian Hacking has pointed out, vindication is a central element in achieving the status of science.[19] Usually this is understood as a phenomenon internal to science, but, seen from the perspective of the epistemic periphery, we are reminded that self-vindication is also something achieved socially and "produced" time and again in the relationships (of scientific conferences, diplomacy, peer review, business proposals, and so on) that circumscribe contemporary scientific forms. While it is logical, therefore, to expect science to be differently articulated wherever you go, it is not always possible to predict how it will be so.

While there is no doubt that the interferon project was a formative part of the emergence of biotechnology in Cuba, and while there may have been very considerable political will behind the project, usually embodied in the figure of Fidel Castro, we have to look elsewhere for our story of biotechnology in Cuba. Indeed, "no one," Michel Foucault counsels us, "is responsible for an emergence, since it always occurs in an interstice."[20] Foucault elaborates on this point: "[An] emergence is the designation of a space, a site where the struggle of forces takes place, a place of differentiation, an arena where struggles for domination which have no 'progressive' aim are realized within the play of dominations."[21] This observation formed the bedrock of his critique of contemporary historiography and his attempt to redefine it through developing what he termed a genealogical mode of analysis: an approach to history that focuses on the absences and ruptures as much as on historical continuities. It is an approach that also provides a suggestive way of accounting for the emergence of biotechnology in Cuba in the absence of the central characters and determining factors that underpin the official historiography of the "event." If we want to get a handle on what this unique space of operations we might label "Cuban biotechnology" became and what became of "it," it is thus precisely to the "struggle of forces" involved in the *making* of Cuban biotechnology that we must turn. These struggles begin, of course, in the imagination.

TWO

# Imagining Science

Just off Calzada del Cerro, a long dusty stretch of road running south of the Plaza de la Revolución, is an unprepossessing two-story building. It looks just like all the others here: crumbling impasto walkways beneath a concrete brown facade. Toward the plaza end is the office of Dr. Gregorio Delgado, one of Havana's foremost historians of public health. Inside, the strip light is bare, netting strung up to catch the halogen bulb should it fall.[1] Here you can peruse a good number of medical journals, testimony to the long association between science and development in Cuba. The island has long been a crossroads between the Old and New Worlds as well as a locus of scientific modernity itself: the first electric railroad in Cuba was built in 1858, and it was in Havana in 1849 that Antonio Meucci discovered the possibility of electronic transmission of the human voice while carrying out electrotherapy on a patient suffering rheumatisms in the head.

During the neocolonial period of 1902 through 1959, when Cuba was under the de facto control of the United States, however, scientific research on the island was constrained. Colonial powers are not renowned for their generous investment in local science. The Cuban revolution, by contrast, has been. And while such attempts to develop a scientific capacity under socialist imperatives prompted historian of science Loren Graham to observe warily of science in the Soviet Union that "[r]evolution is based on discontinuity, but science relies on continuity,"[2] science in revolutionary Cuba had in many respects to start from scratch. Nonetheless, in his assertion that the future of Cuba had to be a future of men of science, Castro was acutely aware that there was an older, more significant relationship between science and the state that the revolutionary government could draw on in constructing that future. These local and historicized framings of scientific practice were reformulated under the revolution against the backdrop of broader global politics.

## Medicine, Myth, and Modernity in Cuba

### *The Dance of the Millions*

You would have to go back as far as the early seventeenth century to find a nationalist discourse in Cuba independent of the claims of science, and at that point nationalist discourse was weak.[3] As soon as calls for Cuban independence from Spain began to be voiced by the likes of José Agustín Caballero (1762–1835), Félix Varela (1787–1853), José de la Luz y Caballero (1800–1862), and Tomás Romay y Chacón (1764–1849), they were tuned to the promise of science. While each of these Cuban independence figures explicitly saw science as a means of achieving a free and modern Cuba, it is José Martí who is most usually accredited with the systematic promotion of science in the name of national development. And Martí, of course, would become not only the founder of Spanish-American literary modernism but the ideological figurehead of the 1959 revolution. It was in the figure of Martí that the twin calls for national independence and scientific modernization coincided in the realm of Cuban politics. And while Martí himself died on horseback leading a charge against the Spanish during the War of Independence, these two discourses would be reconvened under Fidel Castro's revolution.

They would first be put through the mill of American empire, however. The great scientific work of nineteenth-century Cuban science is a case in point. Carlos J. Finlay's 1881 discovery of the *Aedes aegypti* mosquito as the transmission vector for yellow fever was based on his study of the disease as a specifically national and regional problem.[4] More than a little irony was thus appended to the emergence of dengue in 1981, delivered by the same resilient mosquito exactly one hundred years after Finlay submitted his results to the Cuban Academy of Sciences. Amid much controversy and recriminations from both sides of the Florida Straits, Finlay's hypothesis did not gain recognition until the results were confirmed by the United States Army's Walter Reed Commission of 1900.[5] Here, as in many other examples, the connections between science and development were limited by the social dislocations of colonial rule; behind those, in turn, was the manufacture of vast quantities of sugar. For all the emancipatory potential that Cuba's domestic intelligentsia had seen in science, therefore—such as the creation in 1861 of an Academy of Sciences in Havana—the truth of it was that the strongest impetus for its development came from the demands of the increasingly American-owned sugar industry.[6]

Sugar had been the motor of economic growth in Cuba since the eighteenth century. It had provided the basis (culturally, politically, and socially) for a quintessentially Cuban form of modernity: a starkly divided society in which huge resources supported a wealthy elite while the vast majority went without.[7] This was merely exacerbated from 1902 to 1958, when Cuba was under US neocolonial rule. American capital poured in under preferential tariffs, financing a growth in the sugar industry of 54 percent between 1923 and 1929. This fantastical glut of mass production was known as the "dance of the millions." But this growth was accompanied by a decapitalization of national interests, and domestic scientific development was stifled.[8] It was the principal reason why the Truslow Mission, sent to Cuba on behalf of the International Bank for Reconstruction and Development in 1950, found not one adequate research laboratory.[9] Castro's declaration in 1961 of the importance of science was not just a statement of intention, therefore, it was an observation of necessity: now that it was in charge of its own affairs, the Cuban state would have to free itself from the shackles of a sugar economy. But it served another purpose too. In reconnecting the ideology of the revolution with the political ideologues of the nineteenth century, Castro found a ready-made vision of the role of science in the pursuit of an independent national development. It was a vision he found still apposite.

## Magical Modernism

Castro wasn't the only postcolonial leader concerned to develop a scientific base in place of a dependence on agriculture, of course. The policy was even more explicit in Israel, where Theodor Herzl's writings were drawn on in a manner not dissimilar to Martí's, in particular for their invocation of science in overcoming natural limits to the realization of the nation-state. Similarly, Jawaharlal Nehru in India, Gamal Abdel Nasser in Egypt, and the party leadership in China all displayed an obsession with science as a means to achieving national progress.[10] In a statement closely echoing Castro's own portentous declaration, Nehru himself claimed that the future belongs to those who cultivate science and befriend scientists—advice he certainly took note of himself. But with its sugar-dominated economy, Cuba was a classic "nature-exporting society," to use Fernando Coronil's telling phrase. Sugar thus created not only an unequal exchange of value flows (the primary dynamic in most theories of underdevelopment) but involved also "the formation of subjects as much as the production of values."¹¹

Coronil enjoins us to see "the so-called periphery as the site of subaltern modernities, rather than as the region where traditional cultures are embraced by Western progress." In such an account, the "imagined community" of the nation is therefore constructed as much by the materiality of the resources on which that nation is built, and the concomitant forms of development that such material resources make possible, as it is by the ideational realm.¹² For Coronil, certain commodities, such as oil, or sugar, can themselves provide important sets of ideational resources for their perceived ability to convert "land" into "value." Is it possible that biotechnology in Cuba took on something of these "magical" qualities? Certainly it was perceived that science could link the material and ideational realms through an enhanced ability to manipulate that constrictive materiality, to provide the state with new material (and thereby political) resources. These materials would no longer be raw materials, like sugar, of course, but high value-added materials with high political value. If so, this was very much against the trend of things elsewhere: by the 1970s the large-scale purchases of technology undertaken by the developing countries in the 1960s had failed to create any significant benefits, and the oil crises of 1973 and 1979 made it harder still for these countries to develop their science and technology systems.¹³ It was at precisely this point, however, that the development of science in Cuba was accelerating faster than ever.

## *Scientific Socialism*

There is no such thing as one type of capitalism, nor indeed has there ever been one true socialism. Many types of capitalism and many types of socialism have jostled along despite, because of, and alongside one another. Cuban socialism took root in and through the state's attempts to move away from sugar, and the development of Cuban science policy went hand in hand with this broader dynamic of socialist state formation. Soviet Russia offers the obvious benchmark, where science was characterized by the centralization of administration, long-term planning, and the requirement of *vnedrenie*—the assimilation of the achievements of science and technology in the productive process.¹⁴ The biological sciences languished in the Soviet Union, particularly owing to the long-term influence of Trofim Lysenko (lasting from the late 1920s to the mid 1960s), which saw the Communist Party outright reject the function of the gene, and where even the study of genetics was outlawed in 1948. But this is where, historically,

Cuban socialist science has differed from Soviet socialist science. Within Soviet science, the means of scientific research were privileged over the ends because they were themselves considered to be part of the iron laws of Marxism.[15] In Cuba, by contrast, the ends were valued over the means. Any scientific approach would be considered so long as it achieved the required results: there would be no Lysenkoism in Cuban science. The state's influence on science here would take a different path.

Shortly after Castro's declaration of the role of science in Cuban socialist development, Ernesto "Che" Guevara, in assuming the directorship of the Ministry of Industry, outlined a strategy for the immediate resolution of the most pressing socioeconomic problems in the magazine *Nueva Industria Tecnológica*. While the tasks he described were initially modest, the importance of scientific input was soon firmed up. In the mid 1960s, for example, Guevara observed in the same publication that "we need to succeed on the path to development with the most advanced technology . . . and we will have to use technology wherever we can."[16] To which Castro subsequently added, both echoing and reworking Lenin's famous dictum, "today, if you want to produce electricity and don't have the most modern machines available, the cost is that much greater."[17] This would have significant implications for Cuban industry and science, but Castro's comment also alludes to the fact that, while *innovation* was still far from a priority, the need to overcome these sorts of shortages meant that *improvisation* was already a valued ethic among the growing ranks of scientists and technicians. The principal rationale was, as pointed out by Antonio Nuñez-Jímenez, Cuba's preeminent geographer and a personal friend of Castro, to do whatever was possible:

> The Revolutionary Government has plans in which millions of pesos are invested and in which there is a huge factory . . . but not one agronomist; there is nobody to study the earth, [and] there are practically no climatologists. These are plans we are developing without waiting for the ideal conditions to obtain . . . the technology will arise in and through these plans. And we also believe that this sort of practical scientific development will permit us to develop the theoretical base of our country's future science.[18]

Nuñez-Jímenez was an ardent communist and played an important role in ensuring that Cuban science after the revolution became explicitly organized according to socialist principles. Perhaps most notably, this was a socialism that reflected Mao's ideas on creativity and national spirit.[19] By the late 1960s, then, science policy was being described by the state as a necessary part of socialism precisely because it spoke to these national developmentalist policies.[20]

## *Dreaming in Binary*

The central dynamic shaping Cuban science under the revolution, however, has been the dialogue *between* these nationalist and socialist visions of development. As the Cuban scholars Tirso Sáenz and E. G. Capote point out, science policy in revolutionary Cuba has also always implied a concomitant "scientific apparatus": "the full set of measures taken by the government to ensure its realization."[21] This is part euphemism and part expression of what P. Gummett and J. Reppy elsewhere call a "mission-oriented" approach to scientific research, one that is mirrored in France and in the Soviet Union, and that contrasts somewhat to the "diffusion-oriented" approach of Japan or the former West Germany.[22] In many respects the Cuban model of science in fact more nearly approached the French than the Soviet model. Hence, where France had its Grandes Écoles, Cuba had its elite finishing institutions, such as the Lenin-7 and the Humboldt schools. But the *fundamental* difference between science in the Soviet Union and science in Cuba was that rather than science serving socialism, socialism and science were together intended to serve the nation.[23] It was from this vision that the seeds of Cuba's biotechnology project were sown.

The very idea of biotechnology would thereby become both one of the principal goals and perhaps the definitive articulation of a particular form of this rather binary political culture in Cuba. A separate, and rather more officious declaration, made a few years later, outlined one of the main tenets, if not the raison d'être, of the political rationality behind this, a rationality bound up in the socialist reworking of Cuba's national project of modernity: "It is necessary [for us] to seek out the best and the most modern that is possible, but not the best and most modern that is impossible."[24]

It is this headlong pursuit of the possible (but often, also, the improbable) that Cuban scholar Damian Fernandez has diagnosed as *"lo posible."* The promise of *lo posible* has been articulated through two chief discourses in Cuba: a long-standing nationalism and the historically more recent co-option of socialism. As a quintessential manifestation of Cuban modernity, biotechnology had to be amenable to both of these, and it would bear the marks of so being. Such comments relate to what the American sociologist John Dewey saw as the emancipatory possibilities of science. "Ultimately and philosophically," Dewey wrote, "science is the organ of general social progress."[25] The emergence of biotechnology in Cuba seems, initially at least, to fit this diagnosis. Certainly that was the intention of the socialist state when it called for the rapid development of biotechnology in the early 1980s. The improvements in Cuba's health-care system were already being heralded as the "revolution within the revolution," and it was hoped that the biotechnology project might similarly constitute a "scientific revolution" and provide the impetus for national modernization too. Like Dewey, Cuba's new socialist government saw science as giving humans control over nature and thereby also their own destiny. But even if high hopes were attached to the project from the start, they never coalesced into a master plan. The emergence of the biotechnology program was much less specified and remained closely bound up with shifting social and political events. In order to locate that emergence among such events, we need to know something of the particular developments that converged to create the space of possibility where biotechnology was to emerge center stage in the revolution's plans for modernization.

## The Cuban Biotechnological

In 1979, before the idea of developing interferon had been proposed, a group of Latin American countries met in Havana to discuss the priorities in developing biotechnology on the continent. They recognized it as a potentially powerful technology and underscored the need for their governments to adjust the sorts of biotechnological programs the world was then seeing to fit the industrial priorities of their own countries. In Cuba, this would come to be rather more readily achieved. Between the late 1970s and the early 1980s, three characteristically modern aspects of Cuban political culture converged around the issue of national development. Together they would supply precisely this space of opportunity. First, a change of approach within development planning saw an increasing interest in the potential of more scientific forms of development. Second, a crisis in the productive sector (the sugar industry) occurred without a corresponding decline in capital inputs owing to the greater availability of credit from the Soviet Union. Third, the maturation of the country's health and education systems provided a substantial pool of knowledge in medical science. While each of these processes had been unfolding independently and in different ways, they came together in the late 1970s around a discourse of biomodernization: one that institutionalized a particular developmentalist approach to science, centered on the social body and the regime's capacity to intervene in individual health. These processes provided the essential conditions of existence for biotechnology and, in turn, biotechnology provided the means for their fuller articulation. In the process, new spaces were called forth, new organizational forms took shape, new actors emerged, and new policies sought to mold these elements together.

## Planning for Science

For the first two decades of the revolutionary period, developing an advanced scientific sector was not a possibility in Cuba. While there may have been an interest in science, the overwhelming priorities were with the revolution's basic industrial and social programs.[26] But in the late 1970s, development planners at the Junta Central de Planificación (JUCEPLAN, Cuba's central planning board) began looking for alternatives to sugar after the failure of the large-scale industrialization plans made early in the decade.[27] The pinnacle of this failed approach had been the attempt to produce a record 10 million metric ton *zafra* (harvest) in 1970. It was described at the time by British historian Hugh Thomas as a "grand Potemkin-type harvest" whose long-term impacts could not easily be measured. A few short-term consequences were clear, however. First, the state had massively expanded its reach into the planning sphere.[28] Second, development planning switched immediately from a model of extensive natural resource development to one of import substitution industrialization.[29] Third, the problem of modernization was reclassified.[30] Simply put, the aim was now to ensure that, by 2000, the industrial sector and not agriculture would be the major contributor to the balance of payments.[31] The problem, however, was that outside of sugar production the options for industrial development were scarce.

And the situation only worsened. A spiraling hard currency debt (as Cuba persisted in importing investment goods from outside the Council for Mutual Economic Assistance [CMEA], the Soviet trading bloc) and a growing reliance on the CMEA exacerbated a retreat to bureaucracy. Soon every productive unit had its plan, and every plan was part of a larger plan. There was, as a result, little if no capacity for innovation in any Cuban enterprise. Something had to give. At a meeting convened to set out a new development platform the call went out to "better incorporate scientific and technical achievements."[32] It was a call that effectively translated into a search at the highest levels of government for a high-tech showpiece industry. Electronics and nuclear physics were the immediate choices. By the start of the 1980s, Castro's son, Fidel Díaz-Balart, was himself in Moscow, training as a nuclear physicist, and the government was channeling most of its science and technology budget into computers and nuclear energy. But computers and nuclear physics soon proved untenable, not least because the country had no history of scientific development in any related fields. So when Randolph Lee Clark first told Castro of the potential of biotechnology, here was a second opportunity. Biotechnology wasn't cheap but it required a smaller outlay of capital than nuclear physics and used resources the country in large part already had. Small-scale, high-return, and, most important, new to the island, biotechnology was a welcome prospect if it could put some of the "grand plan" ghosts of the past to rest. Hence, while the norm was for development to be organized at the state-enterprise level, when Castro sent his first scientists abroad to learn the basics of genetic engineering he was both sidestepping the planning bureaucracy and embarking on a route that was, paradoxically, both more administratively decentralized and more politically centralized than the norm.

### *Making Ends Meet*

Running alongside these developments in the planning sphere was the development of revolutionary scientific policy. Immediately following the revolution (from 1959 to around 1965), the government sought to crisis-manage the remnants of the country's scientific infrastructure left over from the previous political order. Scientists and foreign technicians were contracted from abroad to help put this infrastructure in place. Beginning in 1965 with the foundation of the National Center for Scientific Research (CENIC), a new scientific infrastructure began to take shape. This period of consolidated scientific planning saw the creation of a research-oriented infrastructure through, for example, the formation of a Deputy Ministry of Teaching and Scientific Research within the Ministry of Public Health (MINSAP) in 1973 and the establishment of the National Council of Science and Technology (CNCT) in 1974.[33] From around 1975 emphasis was then placed on industrialization and application (1975–80) as the previous aims were consolidated into a national scientific policy at the First Party Congress of the Cuban Communist Party (PCC) in 1975. This was the beginning of a process of channeling research into selected key areas. The biological sciences were one such area, and they would come to form a dominant part of CENIC. From the late 1970s—most visibly marked by the space flight of the first (and to date only) Cuban cosmonaut—scientific development was properly considered part of development planning.[34]

This revolutionary science policy articulated three consistent sociopolitical aims: the fruits of scientific research were to be domestically grown; results should be mobilized and made available as soon as possible; and science as a system should be responsive to the more urgent problems of society.³⁵ The enthusiasm with which scientific policy makers would take up biotechnology is at least partly explicable by the way that the interferon project so effectively fulfilled these three principal objectives. The work in House No. 149, for example, was a clear case of technology transfer, but more than this it was a recapitalization of the nation's science: Cuba sought out the know-how in Finland, not the hardware which was largely assembled ad hoc in Cuba (and of course it took expertise from Clark in the United States, as well). Second, Cuban work on interferon in Cuba also involved the repatriation of innovation, because to repeat the Finnish process in Cuba did indeed require bringing together the materials, laboratories, patients, and so on that are required in order to produce interferon. Third, the use of interferon in the dengue fever epidemic provides a clear example of the mobilization of these results into the realm of social practice.

But a dual movement was taking place in all this that would have significant implications for the development of biotechnology in Cuba. On the one hand, the development planners were opened up to an ethic of experimentation (with greater value placed on finding new solutions to old problems). Scientists, on the other hand, became more strongly subject to a bureaucratic urge and the centrality of the plan, with greater value placed on projects which worked within the system to achieve socially and politically useful results. Indicative of the institutionalization of an experimental ethic was the formation of a national movement and later, in 1985, the establishment of an annual Science and Technology Forum from what had previously been informal meetings of scientists, technicians, and workers to recover and produce spare parts. The bureaucratization of science, by the same token, could be witnessed in the coming to prominence of the Cuban Academy of Sciences (ACC). The ACC had become a redundant institution during the US neocolonial period (1902–59). The revolution immediately began to reinvigorate it, first restoring its academic capacity and then extending it into the realm of planning itself. Toward the late 1970s the emergent terrain between development planning and science had therefore begun to be charted through the new policy regime centered on the National Science Policy (Política Científica Nacional) and the National Economic and Social Development Strategy (Estrategía de Desarollo Económico y Social del País). The stance of the ACC at the head of this system was clear: any project that would please both the planners in its novelty, and the scientific bureaucrats by its being easily incorporated within the highest levels of the state, would be well positioned to receive its support.

## Healthy Bodies, Healthy Minds

The advances made since 1959 in health and education are widely recognized as being the two greatest achievements of the Cuban revolution.[36] Cuban medicine before the revolution had been characterized by an advanced level of health care—provided mostly by private practices in urban centers for the elite. There was a dearth of health-care provision for the majority of the population. In 1959, an exodus of medics who were primarily members of the now disenfranchised middle-classes began. By October 1961, 3,000 of the 6,250 medics registered in 1958 had left the country.[37] While this put a great strain on the incoming government to find and train people to take their place, their exit effectively provided a blank canvas for a new medical labor force; medicine became a cause célèbre.[38] It was said that those taking up medicine under the banner of the revolution would be an "intellectual and cultural force" to promote its values.[39]

Like science, health was also conflated with development and the revolution's extensive health programs increasingly pushed Cuba toward a developed-country health profile. The progressive decline of eradicable diseases and the growth of degenerative diseases, caused partly by the prevalence of smoking, the effects of a fat-rich national diet, and exacerbated by an aging population, make this clear.[40] From the late 1970s heart disease and cancer were the principal causes of morbidity in Cuba. There was, therefore, a genuine medical need for biotechnological products such as diagnostic kits and medicines such as streptokinase. But not only did the health system provide a central material condition of existence for biotechnology, it also framed it discursively via its institutionalization of preventive medicine and an outreach focus.[41] By the late 1970s, Cuban health policy was structured around the mobilization of medicine in the pursuit of social health and political capital.[42]

Where biotechnology in Cuba stands apart from the health system, however, is that it is not participatory: it does not necessarily need to involve the population at large. As we have already seen, the development of biotechnology in Cuba was founded in part on the selection of a key group of highly trained individuals, all of whom were deemed exemplary scientists and exemplary citizens. Nevertheless, they in turn had to be drawn from a wider scientific community, and the formation of such a community was something to which the state had given considerable attention. As one member of the ACC and former professor at the teaching hospital Calixto García recalled in 1985, it was through education and in particular through the Cuban education system's scientific focus that medicine would fulfill its function as a principal productive force.⁴³ Immediately following the revolution, the University of Havana was overhauled with sweeping reforms, including a crash program to increase the number of physicians and the introduction of new courses, such as biochemistry, as well as compulsory classes on Marxism for all. In these feverish years, one commentator on Cuban medicine noted a "somewhat cavalier disregard, probably impossible to avoid in the circumstances . . . of certain scientific concerns, methodology of research, and research itself."⁴⁴

It was not just scientific and medical institutions that were overhauled. Such processes also broke down the institutionalization of scientific practice itself. New standards that were more in line with the West were instituted throughout the 1960s, and by the mid 1970s, when control over medical education was passed from the university to MINSAP, spending on education had leapt to twice that of the previous decade.⁴⁵ The scientists who formed the interferon group were drawn from this first generation of scientists to be fully trained under the revolutionary education system, providing what one scientific director described, not unrealistically, as a "critical mass" of biomedically trained scientists.⁴⁶

## Biomodernization

Such comments warrant closer scrutiny, however. The rights to health and education in Cuba have been described as "two sacred conquests which must be held above all others."⁴⁷ Even when this comment was made, at the height of the Special Period—when the country was effectively in economic ruin—health and education continued to be subsidized and made available for all.⁴⁸ That indeed was the message of "Health for All" (Salúd para Todos)—the title of a series of annual conferences in Cuba—which broadcast the message at home and abroad. But it was a message backed up with two major medical-political programs. At home the family doctor system aimed quite literally to put a doctor on every block, while abroad the Cuban government continued to send thousands of young medics on health missions to other poor countries. A similar approach was taken to education, with the founding of the Latin-American School of Medicine for foreign students. It is useful to trace the way such values fed into biotechnology. The Cuban state accrued political capital through its promise of ever better health and educational standards. The Cuban scholar Damian Fernandez argues that a major trope of the revolution has been a paradox of desire and disenchantment. In a discussion of what he terms the "politics of passion," he argues that the revolution has to consistently promise a better future in order to obtain the political capital it needs to remain in power in the present, and yet it has proved consistently incapable of making good on many of those promises.[49] The result is a dialectic between desire and disenchantment which gradually erodes the legitimacy of the government.

In some senses, biotechnology presented itself as a sort of technological fix for this process that took place in the realm of the mythical promise of modernization. First, with the advent of biotechnology, the sites at which that promise of modernization is articulated are transposed from the sugarcane fields to the laboratory, where indeed the promises of modernity look more promising. Of course, this is in part because they are no longer seen—a certain mystification takes place. Second, biotechnology offers a new lease of life to the promises of better health and education that were faltering, ironically, as a result of having been so successful. Since the earliest days of the revolution the decline in infectious diseases had been witnessed in the official press with fastidious detail. But as we have seen, these figures can only go so low, and by the early 1980s Cuba already had a rich-country health profile. On the one hand this provided considerable political capital but, shackled to its own utopianism (the discourse of perpetual improvement), the state needed more evidence of change. The same held for literacy statistics. In this regard, biotechnology provided an apex to both the health and education systems, and that apex provided a space in which they could both be reproduced: biotechnology requires better trained individuals and a more advanced health service. Little wonder then that in 1984 Castro noted that "biotechnology is of enormous social importance . . . by giving attention to such priorities we can continue with our social programs."[50] In short, biotechnology gave these two most crucial realms of symbolic capital in Cuba an added lease on life.

But the ethical imperatives behind Cuba's remarkable achievements in health and education were also mobilized *within* the emergent biotechnology project. Health and education as value-forms are both expressions of what two Cuban authors, E. R. Borroto Cruz and T. R. Medrano, refer to as "socialist humanism."[51] Socialist humanism is an ethic for political conduct within such spheres as science, health, and education, and it is situated within a modernist rationality: it is organized through the bureaucracy and in the name of social welfare. At the same time, in Cuba this ethic is habitually coupled to a form of nationalist pragmatism. "In Cuba we don't have great natural resources," a former investment manager involved in the biomedical sector said; "For this I think the most important thing is to develop medicine. From the social point of view other industries don't have the relevance medicine has." Here then were the means for developing biotechnology devoid of a logic of capital and structured instead around the deployment of a socialist-humanist ethic geared to the necessary realization of specifically national goals. The relationship between these two sets of ethics is at the heart of the discourse of biomodernization.

Toward the end of the period I spent trawling through health documents in Gregorio Delgado's office, he invited me to share my own thoughts on the history of Cuban medicine. The one question that still really lingered for me was cost: why entertain a form of science renowned for being so expensive to get up and running, let alone sustain? His response was:

> The more biotechnology you have the more money you need to finance it: for most of the Third World, to fight against non-infectious diseases is not necessary, but to fight against them is for us because we have eradicated most of the other diseases, and so non-infectious diseases such as cancer and heart disease come to take priority. This is expensive to treat, but if you are committed to everyone, then as these diseases become important you have to fight them. But, when men form together in a state—in which private institutions can function, so long as they are not allowed to keep all the money—this, then, in a sense, is a little like the *Real Sociedad* [colonial organizations founded across Spain and some of the Spanish territories, including Cuba, to promote cultural and economic development]. The money is important, but the human resources are the *principal* resource: health and education *can* develop the economy.

The arguments of a society of gentlemen formed in the seventeenth century to combat such problems as sanitation, civility, and public order had just been given a highly novel twist. In so doing, Delgado also captured the essence of Cuban biomodernization: a set of broadly understood aims that sought to convert the past achievements of the revolution in the scientific (and particularly the medical-scientific) field into a coherent development program for the future.

## Pharmaceutical Dependency

However, the object of biomodernization was not, properly speaking, society. It did not seek to reform the social, to create the "New Man" of Che Guevara's utopia.[52] Rather, it was a product of the post–Great Debate era.

Instead of just attempting to shape society through technology, biomodernization also sought to shape technology through society, making use of its now consolidated scientific infrastructure and skilled personnel. Biomodernization was born of a Cuban capacity for improvisation, then, but it sought to transform this into a comparative advantage based on innovation. In doing so, its aim was to escape the chains of economic dependency on sugar (and, as the next chapter examines, the political dependency on the Soviet Union that came with it). But biomodernization was itself strongly shaped by two formative elements: on the one hand, the political "medicalization" of the Cuban state and, on the other, what we might think of as the economic "medication" of the Cuban state by the United States. What needs to be understood is how together they triggered the turn to biomodernization at the end of the 1970s.

## The Medical State

If science had long formed a part of the project of national modernity, medicine in particular was prominent within the inner circles of the socialist state. Che Guevara was perhaps the most well known high-ranking medic, but other physicians such as Dr. José Miyar Barrueco (known as Chomy) have long had the ear of Castro and were a force to be reckoned with in Cuban politics as much as in Cuban medicine. The medicalization of the Cuban state was in part something that took place at the level of personnel, therefore: scientific elites were also often political elites.[53] But it also took place through the pervasive connection between the national and the individual body in Cuban political culture; through this symbolic connection, public health has become central to the process of state formation.

Two aspects here are relevant to the story of Cuban biotechnology. First, the health system has reached more profoundly into society in Cuba than perhaps anywhere else. This provides the state with a significant means of control of the population. And the massive improvements in public health that have been possible on precisely this basis have also played an important role in fostering legitimacy for the state: the commitment to having a doctor literally on every block, for example, allows for considerable attention to health and also for considerable amounts of data to be obtained about individuals; the policy of medical diplomacy makes a very specific connection between caring for the self (and others) and particular models of (preventive and curative) citizenship. The health of the personal body in Cuba has therefore, in a very profound sense, become a constitutive part of the health of the national body and a model for the organization of society. The revolutionary state's instantiation of this new form of (health-related) governance thus promoted new forms of citizenship that were reflected in, and in part orchestrated by, these new forms of public health policy and practice.

Second, under such a rubric, a significant discursive development has taken place. Contributing to the individual and thereby the nation's health (a central task of socialist humanism) was largely seen, at the start of the 1980s, as synonymous with contributing to the nation's wealth (a central task of nationalist pragmatism). It was in this context that biotechnology offered (temporarily, as it would turn out) a binding together of nationalist and socialist visions of modernization. One of the reasons that health has been such a prized asset in Cuba is precisely because it makes this discursive connection. It offered an appropriate means for the medical state to adapt both to domestic needs and the exigencies of global capital, on the one hand, and to commitments to international socialism, on the other.[54] By extension, one of the reasons biotechnology was so attractive to the Cuban state was because it offered the opportunity to further bind nationalist and socialist visions of development together within a self-consciously articulated notion of the "medical state." Hence, while apparently not even countenanced in the Plan Programática of 1976, biotechnology nevertheless became the quintessential expression of its aims and of the drive for modernization it inaugurated. What was really being witnessed in these developments, then, was the reformation of the state through its attachment to new technological forms. And as part of this medical state, biomodernization came to be seen as an eminently *possible* discourse.

## The Medicated State

But this medicalization of the Cuban state had itself to take place within Cuba's very particular international setting, since the constraints put on the country by the United States embargo effectively shut off the possibility of modernization by many other routes. The embargo was enforced by executive order of John F. Kennedy in February of 1962, and medicine was in fact precisely one of those areas the embargo sought to target. The legal scholar Peter Schwab provides a succinct summary of the ways in which the embargo is, as he suggests, a "war against public health":

> Because most antibiotics are produced under U.S. patents they cannot be exported to Cuba under terms of the embargo. Of the more than 300 ma-

jor drugs on the market since 1970 nearly 50 per cent were of U.S. origin and thus effectively blocked from shipment to Cuba. Additionally, the various embargo acts disallow Cuba from using the U.S. dollar in international trade, costing the country additional money for exchanging currencies. U.S regulations also disallow the re-export of U.S. products from a third country, while products even developed through the use of U.S. technology or design cannot be sold to Cuba. Any domestic corporation, foreign enterprise, or third country found doing so can be slapped with U.S. sanctions, while companies that "traffic" in nationalized property in Cuba (that is, use its facilities) may find their executives prohibited from entering the United States and can be sued in U.S. courts.[55]

What has actually been achieved by the embargo, more than the political changes sought, is considerable demand for domestically produced materials (including pharmaceuticals) and the fashioning of a political arena in which Cuban-American rivalries can play out: both of which lent support to Castro's decision to develop biotechnology, so as to overcome the island's pharmaceutical dependency on other countries. By such means, biomodernization also became an eminently *probable* discourse.

All of this sheds a rather different light on Georges Canguilhem's classic observation that "the past of a present day science is not the same thing as that science in the past."[56] What had been put into place in Cuba in the early 1980s, therefore, was not just the capability to do biotechnology but a realm which both demanded, and made possible, the realization of biotechnology as part of broader social and political processes as much as a specific domain of scientific practice: biomodernization. It was because these structures were in place that Cuban biotechnology, when it emerged, would develop independently of the culture of science then regnant in Euro-American biotechnology circles. The political will of the state may well have been behind it, but the necessary resources could only be put together by the state (admittedly at Castro's behest) because connections had already begun to be made between them in many other seemingly invisible ways.[57] Pharmaceutical dependency and biomodernization thus went hand in hand. While developments within the social body at home meant that individuals had need of the more advanced drugs that were more susceptible to the policing effects of the embargo on the political body of the state, the state's response to this crisis was increasingly skewed toward the one area in which it was coming to have a real comparative advantage: biomedicine. This intersection was the point at which biotechnology emerged. But if, to this moment, biotechnology had taken the form of an unspecified emergence or "fracture" of existing policy and practice in Cuba, it would not remain devoid of attempts to specify it for long. Politics had long formed a part of the culture of science on the island, and this new and still largely untested science was about to become rather more actively taken up within political culture at large.

THREE

# Making Space for Science

At a Sunday lunch party with some friends living in the specially provided apartments opposite the CIGB, we are discussing how work on interferon paved the way for a full-fledged biotechnology industry in Cuba. My hosts, Jorge and Mileidis, are biochemists. Their apartment is on the fifth floor and has views over what seems like a whole suburb of housing units, most of which (apart from the significant minority that sit unfurnished, not to mention unfinished) are inhabited by scientists. Over traditional Cuban fare of rice, pork, and beans, Jorge and Mileidis tell me of their work. A few days later, on a tour of the CIGB, I stop at a small shack somewhere on the grounds where a radio grunts out salsa. Here things that would otherwise be headed for the scrap heap are rehabilitated; bicycles are given a new lease on life and taken back to the scientists' homes. The whole thing seems a little incongruous when set against the BSL-4 labs in the building behind us. But the CIGB is a workplace like any other, and such shacks often exist behind government buildings, hotels, or any other enterprise. The place itself and the scientists that work here have to be maintained.

Historians of science have long paid attention both to the "made" qualities of science and to the importance of context in scientific endeavor: the fact that ideas have to be produced within particular places, that knowledge is embodied in both people and places, and that, to the extent that science travels in the world, it has to be remade everywhere that it goes. The world of the laboratory, insofar as it is indeed a part of the world, thus develops in particular contexts. We saw in the previous chapter that biotechnology offered a means to overcome an endemic problem of government for the Cuban state: the need to service what were often conflicting nationalist and socialist visions of development. But how, physically, did the various actors who then got involved in this project (from the government, to its planners, to the scientists, laborers, and even foreign consultants) manage to make such a vision come to life?

The most obvious answer is that the biotechnology project was a product of the large-scale mobilization that brought these various elements together and helped physically weld them into the longer-running narrative of Cuban development. But also warranting our attention is a more subtle series of developments where—as the biotechnology project made the transition from cottage industry to "big science" in Cuba—the social relations of scientific activity became spatialized in a more profound sense. In the process of all this work, however, the tensions between nationalist and socialist visions of development also became institutionalized. These tensions would turn out to be a very crucial part of Cuba's biotechnology endeavor.

All of this first became apparent during Cuba's candidacy for the proposed United Nations Industrial Development Organization (UNIDO) International Center for Genetic Engineering and Biotechnology between 1982 and 1984, but it became even more highly visible with the opening of Cuba's own Center for Genetic Engineering and Biotechnology (CIGB) in 1986. With the opening of the CIGB, biotechnology was officially upscaled to a prioritized industrial sector, and the socialist elements of this work were emphasized. But this scientific achievement coincided with the announcement of a political "rethink" known in Cuba as "Rectification of Past Errors and Negative Tendencies," essentially a return to the more nationalist roots of the revolution. In 1986, then, the socialist and nationalist discourses of science that had begun to cast shadows on each other during the UNIDO competition were both bound tightly to the very sinews of biotechnological work on the back of an emergent geographical order composed of a new political discourse (the policy of Rectification) and the spatial expression of that discourse in the work of the agents involved as well as in the organizational, institutional, and social forms of what was by now a rapidly expanding biotechnology project.

## Agents: Pedro López-Saura and Luis Herrera

### *The UNIDO Competition (1982–84)*

After the initial successes with interferon in the early 1980s, a unique opportunity arose for Cuba to push forward its biotechnology program. It was still in vogue to think of the so-called green revolution as a viable solution to the development impasse holding back the progress of so many underdeveloped countries, and there was considerable international interest in developing biotechnology programs across the global South. In 1981 UNIDO announced a competition for an internationally funded project that would foster biotechnology in the third world. Immediately it seemed a promising option to the Cubans who now wanted to expand their work into new areas, including agricultural biotechnology. From the point of view of Castro's government, an internationally funded center would not only provide the means to do this; it would also be a major political coup. Cuba entered the competition in 1982.

The specific aim of UNIDO's International Center for Genetic Engineering and Biotechnology (ICGEB) was to facilitate North-South cooperation in high-technology science and to act as a conduit for knowledge transfer: it was a landmark project for developing world science. But where to locate the center? In addition to Cuba, over fifteen other countries put themselves forward, including Argentina, Canada, Spain, Yugoslavia, Belgium, Egypt, Thailand, India, and Italy. In charge of the committee overseeing the Cuban application, and therefore directly involved in this process, was Pedro López-Saura, now vice director of the CIGB.

## Pedro López-Saura

After a relatively short wait, Pedro López-Saura, director of clinical trials at the CIGB, ushers me through the reception area of his internationally renowned institution and into a further small waiting area. We order coffee and juice and I set out the thrust of my questions—how does he remember Cuba's involvement in the project? At the point of answering, López-Saura is called off to attend to some small crisis somewhere in the institution's vast acreage of offices, labs, and yet more meeting rooms.

López-Saura obtained a PhD in biological science in 1978 while working in Paris under Nobel Prize–winner Dr. Christian de Duve. He was one of those who had gone to Cantell's laboratories in Finland. Today's crisis resolved, on his return we discuss the initial work on interferon, but the conversation segues into a discussion of how the UNIDO competition interacted with Cuba's nascent biotechnology efforts and, as a result, gave them a certain form. "As we were beginning work in the field of genetic engineering then," López-Saura said, "we decided to make a bid for being the seat of the international center."

PLS: An international commission visited all the countries and evaluated each of them. Then in 1984, a preparatory committee began to meet in Vienna to decide which of the countries would be the seat. But the committee didn't take into account the results of the commission. What I mean is that the committee had made a recommendation of a country. It wasn't Cuba, or any of the selected countries in fact.

SRH: On what grounds do you believe the decision was made then?

PLS: There were basically two opposing groups. One wanted to fragment [*atomizar*] the center, into lots of small parts, so that each country had one part. That group consisted of Spain, Egypt, Morocco, Pakistan. We were in the other group, which wanted a completely integrated center in one place. This group included all the Latin American countries, India, Italy, and a few others. All the countries in our group were backing either India or Italy—we were backing India, which then held the presidency of the nonaligned countries and had a pro–third world [*tercermundista*] position. So we renounced our application in favor of India, as did the other Latin American countries. Then India and Italy united their applications [and] won [by] a very marginal vote. So now what you have is a center in India, one in Italy, and a network of other affiliated institutions. As a result Cuba decided to build this center [the CIGB] as a national center, it was built more rapidly than the international one.[1]

Throughout our conversation, López-Saura dismisses the debates over the location of the ICGEB as being both entirely political and uninformed. As I later find out, his objections to the way the competition was organized can be corroborated with international reports, but what immediately stood out from his comments was his use of the word *aislados*.[2] It is a term that conjures up a sense of geographical as much as intellectual isolation. It points most of all to the connection between physical and epistemological distance. Sociologist of science Harry Collins talks of the "enchantment" of distance in terms of the simplification that summary and dissemination require of scientific practice.[3] López-Saura's comments reveal the contestations that take place around such localizations. Couched in an explicitly socialist vocabulary, they refer again to that important, if paradoxical, trend: namely, that the supposedly socialist objective of the biotechnology project was in fact to provide for specifically national needs.

As a number of other Cuban scientists also suggested, the fact that, wherever it was to be located, the UNIDO center was to be focused on international needs was far more important for the Cuban delegation than the political wrangling for which the competition was broadly criticized. One commentator has rightly pointed out that "although the international centers that promote biotechnology research in the Third World do so with good intentions, the results of their endeavors redound to the benefit of transnational corporations rather than to the peoples of the Third World."[4] So not without reason then did the Cubans suspect that the center would bring no real benefits to the host country. In fact, of the two seats finally chosen (Delhi and Trieste), only the center in Delhi was commissioned to deal with developing country projects, and even as late as 1990 neither of these two centers—which some have referred to as the white elephants of the biotechnology world—was yet operational.[5]

For the Cubans, López-Saura points out, it was clear that national needs would never be met within a framework designed and operated by the advanced industrial nations. Reading between the lines, it appears that such national concerns were articulated through a socialist anti-imperialism (*tercermundismo*). Accordingly, the Cubans threw their support behind the Indian bid—India being a prominent member of the nonaligned countries—and got on with replicating the idea of the center at home according to their own needs and desires.

### *Building on Interferon: The Growth of Domestic Biotechnology*

The emphasis on national priorities that López-Saura alludes to was reflected in the early 1980s in the scientific work of Cuban biotechnologists themselves. If interferon had been produced successfully, it was argued, then it could also be used as a base for developing more advanced skills such as modern immunology, predictive molecular virology, and in particular recombinant technologies (genetic engineering). Biotech companies in the West frequently "spun out" from single-technology platforms into a realm of other applications: witness the growth of firms like Genentech, which in the late 1970s began work solely on the human protein somatostatin and by 2007 had products catering to everything from rheumatoid arthritis to psoriasis. Why not do the same in Cuba?

So new work began on monoclonal antibodies for use in operations and on epidermal growth factors for treatment of skin cancers, another significant health problem in Cuba. But all of this new work had a unifying theme. "The idea behind biotechnology in the early 1980s was to translate scientific knowledge into products which could be used," said a former health official. For a country that until the end of 1981 was still conducting biotechnology out of a converted villa, this drew not inconsiderable acclaim from some, who applauded the government's "bold" commitment to science, technology, and the new informational economy. But it drew barely concealed skepticism from others, who snorted that such a "rash" project in a small, socialist, developing country would be unable even to acquire the infrastructure, let alone survive, in the dynamic and quintessentially capitalist world of the then nascent biotechnology industry.

But the Cubans seemed to have found a way. Indeed, the most immediate problem confronting the Cubans from 1982 onward was how to accommodate the rapidly expanding scale of their work. A converted villa was no longer sufficient. By the end of the year, therefore, a new and larger laboratory was built on the site next door to House No. 149. The researchers moved there in January 1982, and around the more substantial and better-equipped laboratory was established the Center for Biological Studies (CIB), this time consisting of around eighty researchers. Pedro López-Saura was nominated director of this center. As he recalls, this increase in the scale of work itself brooked new approaches to the science:

SRH: Can you explain how the emphasis of work changed, how you moved beyond the initial work on interferon?

PLS: Interferon served for all our methodological development: the first cloning, the first expression, the first fermentation, but then we applied it elsewhere, in growth factor, or the hepatitis B vaccine. . . . We continued exchanges with Cantell for two or three years more, after our first meeting, because Cantell retired and we had changed our interests a little. They were now directed at wider issues, above all the question of genetic engineering, and Cantell's center didn't work on genetic engineering. We retained a close friendship with Cantell, but not scientific relations.

Genetic engineering was necessary, then, because nothing else could meet the demand for raw materials on which to work or for the number of assays or repeat experiments needing to be done. It was chosen not, as in many cases in the West, because it was seen as somehow cutting edge or designer science. It was chosen because it got the job done. But the wholesale shift toward genetic engineering was also, it warrants noting, further indication of the biotechnology project's growing responsiveness to national requirements. Castro wanted to employ interferon as a major clinical focus of medicine in the country and, whatever the wisdom of that vision, Cantell's purification technique provided insufficient quantities to achieve it. José de la Fuente, former vice director of the CIGB and López-Saura's predecessor, recalls:

. . . the jump [to biotechnology] was tremendous. It was mainly MDs sent to Cantell's labs in Finland, with some basic biology. . . . At that time only one of them had some insights into genetics so by the time that the project developed from that initial interferon production to the CIB in 1982, some of them had to go abroad again to learn the science [of genetic engineering].

So while work on purification continued for a while, 1982 saw the instantiation of a parallel project that would attempt to clone interferon. Few others had yet done this. A meeting of the CIB decided to bring in molecular biologist Luis Herrera.

### Luis Herrera

Luis Saturnino Herrera was an outstanding researcher and an expert in molecular biology. He was based at CENIC as head of its microbial genetics department when he was drafted into the interferon team. He had trained in France, Italy, Czechoslovakia, and later worked in Zurich and at Harvard. At that time Herrera was in charge of one group at CENIC that was working on yeasts, while another group was working on molecular biology. Together these groups would form the embryo of Cuba's move into a genetic engineering phase. Herrera was immediately sent abroad, this time to the Pasteur Institute in Paris, to work on learning the techniques involved in cloning proteins. He returned to Cuba in 1983 and was put in charge of the group assigned to trying to obtain recombinant interferon. They succeeded in 1984, and by the time of Cantell's visit to the island in 1986, the Cubans had developed a whole new approach from that based on Cantell's technique: a second-generation interferon, recombinant alpha-2b, cloned in yeast.

Both Herrera and López-Saura combine an intellect for research with an ability to play the diverse roles required of prominent scientists in Cuba; they may not think of themselves as embodiments of Che Guevara's New Man ideal, but their official conduct necessitates a certain recognition of the value of sacrifice and dedication to work. More important, their biographies also tell us a lot about the nature of Cuban scientific work at the time. Men of humble backgrounds, their generation was the first to pass through the Literacy Campaign, the military mobilizations, and the ideological struggles of the 1960s. They were also the first to be fully educated under the revolution. Both men were involved in the biotechnology project more or less from the outset, and they rose to be directors of the CIGB and CIB respectively, before López-Saura moved to join Herrera as vice director at the CIGB.

As Cuban biotechnology research expanded in scope and scale during the period 1982–86, López-Saura and Herrera were also under considerable pressure to commercialize the work that had been done on interferon.

At one point during this period, MediCuba, the organization responsible for commercializing Cuba's biotechnology products, was advertising recombinant interferon before the product had been fully tested. The cautionary response to this by a number of Cuban scientists also indicated an emerging tension between how they would have ideally liked to have proceeded with research and the necessities imposed on them by the state.[6] For now this tension remained muted; the necessities, however, were rather more acute.

## Forms: Rectification and the Biological Front

What then of those necessities? One of the most interesting aspects of the development of biotechnology in Cuba has to be the way in which "the grand plan with no master plan" came to articulate a more general (geo)political shift that Cuba was then undergoing: the process of Rectification. Known in full as the Rectification of Past Errors and Negative Tendencies, Rectification was in effect a realignment of the state's relationship to socialism. More specifically, it was the form in which a long-standing tension between Cuban and Soviet approaches to socialism, and to how they saw themselves vis-à-vis the capitalist West, came to a head. While Cuba's commitment to socialism was never in doubt, what form that socialism ought to take increasingly was. Rectification was the institutional working out of this question, and it sought to upgrade the importance of specifically Cuban values in determining its relationship with the two blocs. Rectification most evidently burst onto the scene in September 1985, when Cuba held seven international meetings over four months to protest the arms race and its consequences for the third world.[7] Here was an explicit orientation of political discourse away from the Soviet Union and toward national needs and priorities. Here too was the means by which Cuban biotechnology would emerge as a thoroughly political form.

### *1986, Part 1: Rectification*

Rectification—officially launched at the Third Party Congress of 1986—was a response to a number of convergent developments: an underlying weakness and inefficiency in government policy that was thrown into relief by a sudden fall in the price of oil and sugar which coincided with rising interest rates; the launching of the Reagan administration's Caribbean Basin strategy, which sought to even further isolate Cuba; and growing pressure for glasnost and perestroika in the Soviet Union, which threatened an economic liberalization not acceptable to the Cuban regime. Triggering these other causes was a growing crisis of party legitimacy: a new party elite and bureaucratic class had levered more of the material rewards accruing from membership in the socialist bloc. Rectification was, therefore, as Cuban scholar Antoni Kapcia explains, "a renewal of old tensions in a new context of economic and political crisis, internal and external."[8] The state argued that the former planning and management system—so central to the nuts and bolts of socialist political economy—was wrongly based on the experience of the Soviet Union and other European socialist countries, and was weighed down from the beginning by concepts and formulas that proved to be inefficient, even in the countries that had developed them. Furthermore, "political work was ignored; the role of ideology was played down; the party's function of political leadership and control was undervalued."[9] As Kapcia goes on to note with hindsight:

> The real crisis point for Cuba was not 1989 or 1991, as it seemed, but 1984–5; once the cycle of debate and contest was under way from that date, the adjustment to the supplementary crises (including the fall of the Soviet Union), although more urgent and more extreme than any had imagined, was actually just that—adjustment of the existing strategy, and not so much a fundamental revision.[10]

Precisely what that strategy implied could not have been clearer: namely a *return* to the "experiences and traditions" of the Cuban revolution, with a greater emphasis on national realities, and on "revitalizing Cuba's *original, creative interpretations* of socialism and Marxism-Leninism."[11]

Both the timing and the content of these developments had a direct bearing on the emergent biotechnology program. By the mid 1980s, new strategies, new objectives, new forms of symbolic and material capital, and a new approach to the pressing questions of development and Cuba's position in a changing world were demanded; biotechnology would become one of the principal areas in which solutions to these questions were first floated. At the same time, the tensions and paradoxes inherent to Rectification would gain institutional and geographical expression in the biotechnology project. If dominant visions of biotechnology's spatiality are thus often conjured up under the rubric of terms such as "space of flows"—a notion which conveys a sense of the inherent transferability of such high-technology universal forms of knowledge—it is worth remembering that such a space of flows nevertheless calls forth a compelling machinery of fixity.[12] At the center of such machinery in Cuba was the Biological Front:

an institution that would give these policies and tensions a particularly geographical twist.

## *The Work of the Biological Front*

The Biological Front was a very self-conscious institutional innovation—a "new organizational form," as official statements had it—and the sort of thing Cuba has a long history of producing.[13] It was established shortly after the return of the interferon group from Helsinki in 1981 as a highly enfranchised scientific body, with policy making powers and direct access to the State Council. This meant it could operate across the various scales of political hierarchy in Cuba in order to coordinate the further development of biotechnology. To the extent that it had its own agenda, this was to insert biotechnology "into social practice" and to "discuss innovative proposals and calculate the human and material resources needed to put them into practice."[14] We might want to specify, however, in what ways, and to what extent exactly, the Biological Front dealt in "innovative proposals." It was certainly not unusual for a developing country to seek to channel the work of biotechnology into practical solutions to pressing problems, nor were the areas in which the Biological Front focused these efforts out of the ordinary: the production of medicines, resolving the country's food shortages, and developing disease resistant varieties of plants. But how best to do this? As the Biological Front set about answering these questions, it quite literally put into place some of the characteristic traits of Cuban biotechnology.

The first head of the Biological Front was Dr. Wilfredo Torres Yribar, a former director of the National Center for Scientific Research (CENIC) and a former president of the Cuban Academy of Sciences.[15] He would be succeeded by another former president of the Cuban Academy of Sciences and member of the Council of Ministers, Rosa Elena Simeón. She would go on to head the Ministry of Science, Technology, and the Environment (CITMA, formed in 1994), the ministry to which the Biological Front is directly accountable. With direct access to the central state, the Biological Front therefore had more immediate access both to funds and political power than would its counterparts in the West. This close association of political with scientific decision making is a key feature (though not necessarily unique if one thinks of the infamous Microgenesys trials of the mid 1990s) of biotechnology in Cuba. But the Biological Front was rather more unusual for the speed at which it undertook the establishment of what, by many standards, was an extensive biotechnological infrastructure. Between 1981 and 1989 the Biological Front poured $1 billion into the rapid establishment of an impressive biotechnological infrastructure, involving a range of research and production facilities, processing plants and supporting units, and the equipment, know-how, and raw materials necessary for a far broader biotechnology program than had been countenanced in the interferon project.[16] Toward the end of this period, the Biological Front would dissolve into a more direct and informal organization structured around what is still known as the Scientific Council.

In order to develop this critical mass of scientific institutes at speed, and despite the obvious constraints facing a poorer country, the Biological Front was also marked by a willingness to improvise. Where necessary, this translated into reworking some of Cuba's already existing scientific infrastructure. CENIC, for example, had been around since the mid 1960s but was incorporated under the Biological Front as an integral part of its overall vision. CENIC not only supplied a number of the initial members of the biotechnology project—both López-Saura and Herrera had been working at CENIC before joining the interferon team—it also fostered an emphasis on the value of basic research within an applied framework.[17] Across the board, institutes were revised and updated with biotechnology production in mind. Or they were simply built from scratch. They pertained in different cases to different government ministries. But in every case, these institutes were woven into a coherent network overseen by the Biological Front.

The Biological Front made use of every available means at its disposal, including new architectural forms. If one looks at the buildings built before and after the Biological Front, the difference is striking. While earlier institutes such as CENIC were ostentatiously modern and futuristic, the institutes created under the Biological Front played down such aesthetic elements in favor of functionality. The new Center for Animal Health, for example, was made from a system of prefabricated units, known as the SMAC system, developed by Cuban architects in order to meet the revolution's needs for cheap and available housing. This architecture was functional and pragmatic, providing easily assembled buildings with large interiors in which laboratories could be easily and flexibly incorporated with spaces for socializing and communicating with other workers.

From as early as 1981 there had been a trend in Cuban science toward "integrating research, teaching, and production."[18] The Biological Front took this a stage further. It aimed to produce centers which were both vertically and horizontally integrated: education, research, development, and production would all take place under the same roof—what the Cubans refer to as full-cycle production. In essence, full-cycle production refers to the linking of upstream research with downstream clinical application and marketing of the products. While this has been taken up in the strategies of Western high-technology firms, such an approach has a particular resonance in a socialist society where downstream marketing in fact refers to social need, and the extent to which information from clinical trials can be fed into the clinical development of a product has considerably greater scope. I asked one project director what full-cycle production meant, exactly. "That everything is integrated," he said. "Here we have research, development, [and] production to the extent of a commercial team."

But full-cycle production only made sense within projects that were sufficiently large to require extensive collaboration, communication, and thereby inculcation, and the projects that most readily displayed a need for all those elements were national projects focused on a guaranteed domestic demand. In short, the scientific infrastructure that the Biological Front was putting together was closely bound up with the national priorities of Rectification.

By the mid 1980s a very specific scientific milieu was thus clearly in the making. But it had yet to be fully realized; for that to happen biotechnology would have to take on a higher profile. The final task that the Biological Front set itself, therefore, was to make biotechnology appear a necessary and, given Cuba's political culture, a logical part of Cuban modernization.

The solution to this problem came with the idea to build the vast Center for Genetic Engineering and Biotechnology (CIGB) in Cuba. Cuba's withdrawal from the ICGEB competition opened up the possibility of developing a similarly scaled institution completely independently. Like the ICGEB, Cuba's own CIGB was envisioned as an impressively scaled, multidivisional research and production space, encompassing over one thousand scientists working with the very latest equipment and set to the aims of national development. It was to cover the full range of biotechnological specialties of the time.

By late 1984—just in time for Castro to reaffirm the state's commitment to biotechnology at the closing session of the National Assembly of People's Power—work on the CIGB had begun. "Our center is being built as rapidly as possible," the delegates were told. What this meant in practice, as one of the builders involved in the project said to the state newspaper *Granma*, was a mad scramble. It would be completed after twenty-two months of working twenty-four hours a day.[19] In 1986 the core of researchers at the CIB moved again, this time into the gleaming new multidivisional space of a laboratory that was then one of the largest biotechnology installations in the world. The CIGB was then, and remains today, the institutional pinnacle in this story of homespun biotechnological development. The application of interferon on dengue patients had been a first turning point, but this was a second and arguably more important one.

## Places: The Center for Genetic Engineering and Biotechnology

### *1986, Part 2: At the Doors of the Future*

The CIGB was opened in July 1986, around the same time that the process of Rectification was also reaching its peak. The CIGB had cost around $25–26 million; at the time the same type of institution might have cost around $250 million in the United States. In a very practical sense, then, the theme for its operation was set from the start: low-cost, high return biotechnology. It appeared to be the answer to what the clamor around biotechnology had demanded: import substitution via the further expansion of domestic drug production. With a further investment of around $100 million, it was fully equipped for research, particularly in pharmaceuticals, immunodiagnostics, and vaccines, while also covering animal, plant, and industrial biotechnology. The CIGB was the first institute in Cuba to be devised *specifically* with full-cycle production in mind. Its internal structure consisted of small research groups, each working on a particular theme: proteins and hormones, vaccines and diagnostic kits, hybrids and model animals, energy and biomass, plants and fertilizers, cell genetics and advanced organisms, restriction enzymes and modification, analytical units, pilot plants, and quality control. Its equipment was widely acknowledged to be of the highest standard, and included mass spectrometers, automatic sequencers, electron microscopes, fermenters, DNA synthesizers, and electrophoresis technology.

Earlier I described Castro's hopes for the sort of scientific miracles the new biotechnology center might presage: the reproduction of the famous cow, Ubre Blanca, for example. But with the opening of the CIGB and the expanded scale of operations it represented, Cuban biotechnology became a "big science" properly speaking, a player on the world stage, rather than a domestic homegrown affair. Accordingly it also became a site onto which were projected the state's desires for not inconsiderable political miracles as well. Explicitly framed in the language of public good, biotechnology now became inextricably linked with state practice: a "vanguard technology" not only *of* science but also *for* the state.

These were the reasons that, publicly, much emphasis was placed on the role of science in society and the importance of scientific community as a means of fostering it. This sort of thing had happened before with state-sponsored discourses around public health. Now the Cuban government was going a stage further and canvassing specifically for the political efficacy of high-tech science itself.

This step had rather more specific ramifications because, for Castro, the biotechnology project in Cuba was a spatial and social project at the same time, and the CIGB was its embodiment. It was spatial in the sense that the CIGB was intended to function as a flagship institution of a mutually supportive scientific cluster. It was social in the sense that good science would not only serve society *but must itself embody* the central components of that society. The two sets of claims were inextricably linked: the social requirements of good science would be more easily achieved because of the particular spatial organization they were based on (the scientific community living together in one place). At the same time as the CIGB was being built, so too was accommodation for the scientists—the same blocks where Jorge and Mileidis now lived.

At our Sunday lunch party opposite the CIGB, it was easy to see the consequences of this arrangement. Jorge and Mileidis are not the only couple here to have met while at work. In a rather more formal register, some of the links between these elements were explored back at the beginning of the CIGB's operation by two rather different commentators: Manuel Limonta, arguably the central figure in Cuban biotechnology at the time, and Albert Sasson, an international biotechnology expert who, through his regular visits to the island and his involvement in the CIGB journal's board of editors, was an active, albeit independent, participant in Cuban biotechnology.

## Manuel Limonta

Shortly before the official political opening of the CIGB, a rather less proselytized scientific event had been convened. It was geared around establishing an international forum for interferon studies (Symposio Internacional de Interferón) and a Biotechnology Society of Ibero-American Countries (Sociedad Iberoamericana de Biotecnología) and was attended by 567 delegates from around the world. Considerable interest had been elicited if only because a number of actual results on interferon's clinical application were to be presented at the Cuban conference at a time when interferon was not yet being clinically applied in the United States. Hence, whereas the opening of the CIGB was intended to sound the political bells of Cuba's arrival on the scientific scene, this event more quietly undertook the same task through the (supposedly) more measured forms of scientific protocol. Among the plenary discussions and speeches at the formal opening of the event, the soon-to-be-director of the CIGB, Manuel Limonta, set the tone with his pronouncement that "we are already at the doors of the future, or what we prefer to call, the future-present."[20] Lest the delegates prove disbelieving, they would be shown around the recently completed facilities later in the conference, many of them subsequently noting that the techniques, organizational skills, physical requirements, and knowledge of biotechnology were evident, and in abundance. The CIGB, the delegates were told, had leveled the playing field between first world and third world science.

Limonta's belief in the future-present was based on the perceived links between some of the central elements of Cuban biotechnology. First was the development of an institution such as the CIGB itself, which brought together, in one place, a concentration of highly trained workers and that in so doing embodied certain characteristics of Cuban society more generally.[21] The CIGB was thus to be both a symbol and a means by which the claims of biotechnology would be realized and Castro's vision of a new scientific practice put into place. The lessons of Cuban society were about to be applied to the laboratory. Nothing approaching this had taken place at the inauguration of the Salk Institute, say, in California, or at the National Institute of Health in Maryland. Second, Limonta believed that the CIGB would provide the impetus for further scientific development as well as a perpetuation of that ordering vision of scientific practice throughout the country.[22] In this way, the CIGB was intended to work as a catalyst across a spectrum of related disciplines, from biology to chemical engineering to physics and, additionally, beyond the sciences, in terms of the physical, social, and political infrastructure that would be required to support it. Biotechnology was to be a forefront scientific sector and a motor for socioeconomic growth. Third, it was to be a center for specifically national development. There was a rather immediate context for this. The Integrated Program of Scientific and Technical Development (Programa Integral del Progreso Científico Técnico, or PIPCT) called for technology transfer to Cuba from within the socialist orbit. PIPCT was approved at the forty-first session of the member countries of the CMEA in December 1985 and was to become operational within the five-year plan for 1986–90. There were five prioritized branches of this program: electronics, mechanical industry, atomic energy, new materials, and biotechnology. It was, in effect, the socialist response to the growing specialization of production in the new post-Fordist world in the West, and was intended to foster an International Division of Labor within the socialist countries. That would never come to fruition, but PIPCT delineated an important component of the way biotechnology was being put to work in Cuba: through establishing a collective of countries that did not recognize Western intellectual property law. At the same time as they gained certain protections from this, however, the Cuban scientists were also looking to expand their own scientific relations with the West.[23] The "future-present" that Limonta claimed it offered was a powerful vision and increasingly, it seemed, a powerful reality. For Limonta, the CIGB would serve to intensify existing patterns of socioeconomic development in Cuba.

## Albert Sasson

Albert Sasson was Moroccan by birth and had pursued a career in microbiology after graduating from the University of Paris in 1967. He joined UNESCO in 1974. Sasson had already published widely on the issue of biotechnology in developing countries when he attended the inauguration of the CIGB, an event that he saw as a major breakthrough for non-Western biotechnology.[24] He advised that "Cuba strongly bear in mind its social, economic and cultural development priorities" as it developed this new tool, suggesting it was not only Castro's government that saw biotechnology as central to development policy more generally. In a statement that mirrored the tone of his subsequent book, *The Challenge of Biotechnology*, he said at the opening of the CIGB:

> The challenge is also, I suggest, to be at the same time a center of advanced research, a center of coordination and relation between research groups, a center of instigating scientific and technical development, and a privileged place in which to scale biotechnological processes up to industrial level.

As for Limonta, the themes of scientific practice, purpose, and community arose. The way they were linked for Sasson, however, was slightly different. For him, the way that the social and the spatial had been brought together in the CIGB allowed for the creative power of new and unforeseen possibilities. Sasson's vision was based less on the assumption of a differentiated global space, as was Limonta's, and more on the assumption that all the elements needed to do contemporary biotechnology science needed to be brought together on one site and made meaningful within a specifically national context. As it turned out, a blend of both visions was to prove the most apposite.

Both Sasson and Limonta concurred that the CIGB provided the Biological Front with the high-profile cornerstone for Cuba's biotechnology endeavors that it needed to make its way in the fiercely competitive world of global science. The most profound achievement of the Biological Front, however, was to be its realization by the late 1980s of a broadly envisioned scientific region: the Western Havana Scientific Pole. It was in this Science Pole—known colloquially as Cuba's "Science City"—that the social and spatial aspects of biotechnology would come together in something approximating a halfway point between these two visions. The CIGB would merely be the centerpiece for this yet more ambitious development.

## Spaces: The Western Havana Scientific Pole (1986–91)

As one of Cuba's most important scientists, Agustín Lage, would write several years later, the development of biotechnology after the CIGB "extended with a [further] round of significant state investment" to eventually include a total of forty-two institutes working on biotechnology research, production, or related services.[25] The Biological Front continued to oversee this work but was increasingly repositioned as the organizational epicenter of a scientific agglomeration fast taking on its own identity. In February 1991, the identity of this growing mass of research centers, production units, hospitals, biomaterial suppliers, and educational institutes, now firmly clustered together in the Cubanacán region of western Havana, was formally recognized as the Western Havana Scientific Pole (el polo científico del oueste). In its physical layout and scope, this geographical demarcation occupied a territory somewhere between a science park and an industrial region. But the "Science Pole" was not the equivalent of such "science cities" as Tsukuba in Japan, or the "science parks" of Cambridge, in the United Kingdom, or Austin, Texas, in the United States. It is worth specifying therefore how the Science Pole does and does not differ from other such scientific agglomerations. While broadly comparable in size, the Science Pole was more clearly thematic (focused on the biosciences) and explicit in its attempt to incorporate science into national economic planning, and not just regional growth or employment creation. On a symbolic level, its location was also significant. The marshes have an important cultural significance within the imagined geography of the city, associating the Science Pole with both the reclamation of nature and the incorporation of the past into a modern Cuban future. If the landscape can be read as a text, this, one could say, was a significant revision.[26]

The Science Pole was further explicitly located at the apex of the two principal axes of urban growth outlined in the Plan Director for Havana, formulated in the 1960s and 1970s: the westward expansion of Havana and the lateral expansion of the new airport and University City (Ciudad

Universitaria) axis. Havana had developed as a typical "monocentric" Latin American city with a historic center from which growth radiated out. Under the Plan Director, new clusters of "creative work" were intended to break down this pattern. A transverse axis would cut across the radial extensions and provide transport and communication links. Three themed "poles"— the clusters of creative work—were planned along it: the Humanities Pole of the University City, centered around the Juan Antonio Echeverría Scientific Polytechnic Institute (ISPJAE), the Industry Pole, and the biotechnology or Science Pole. But as Castro himself pointed out, this was only the beginning: "what began as a relationship based on geographic location became a relationship based on fields of work."[27] Precisely what he had in mind was the promotion of scientific cooperation *between* centers.

But the rationale behind this was not quite, as it would later be in Japan, to turbocharge scientific proximity, collaboration, communication, and creativity.[28] Nor was it designed to kick-start regional growth in the way that science parks generally are in Europe or America.[29] Indeed, being in a socialist state, the Science Pole was bereft of the influence of private capital and property speculation that are determining factors in Western science parks.[30] As anthropologist James Dearing notes in his study of them, the logic of science parks is, rather, to bring together resources and researchers. He notes, however, that "the extent to which a researcher is intrinsically motivated to pursue a task [still] determines what will actually be accomplished."[31] This indeed is the key, and if anything motivation was promoted in the Science Pole more than the need for resources. It was precisely a vision of scientific work that might foster, or otherwise emplace, such an ethic. The form that such an ethic took was modeled on the sorts of informal relationships that Cubans under the revolution had long engaged in: informal, reciprocal, and inventive.

Here the Cubans began to move beyond Weber's diagnosis. For his famous lecture on medicine as a vocation says nothing about what constitutes the good life with respect to work. There might be many answers to such a question: what works in one place is not necessarily what is needed elsewhere. But the answer given to this question in the Science Pole was direction and cooperation: the means of channeling the informal relationships already at the heart of scientific practice. And owing to the concomitant need that biotechnology serve politics as much as it serve science, this would prove to be achievable in rather practical ways. The head of the Science Pole, for example, is Dr. José Miyar Barrueco (Chomy), who is also the secretary of the Council of State and one of the closest figures to Castro. The Science Pole's monthly meetings have usually been attended by Rosa

Elena Simeón (the former minister of CITMA; she died in 2004), Wilfredo López Rodríguez (a government minister and director of Fidel Castro's Support Group), Dr. Concepción Campa (a member of the Political Bureau of the Party, and director of the Finlay Institute), and Dr. Agustín Lage (director of the Center for Molecular Immunology [CIM], and older brother of vice president Carlos Lage). Campa's husband was himself the director of the National Center for the Control of Medicines from that institution's founding in 1989. By similar turns, the president of the Cuban Academy of Sciences (ACC) is married to the deputy minister for the environment at CITMA, whose former superior was Rosa Elena Simeón, and who in addition to being the current head minister of CITMA was the former president of the Cuban Academy of Sciences. Biotechnology may have become "big science," but Cuba remained a small island.

Reflecting on just these relationships between scientists and the state (though admittedly in a context where scientists are less obviously embedded within the state themselves), anthropologist Chandra Mukerji has argued that science always faces a trade-off between autonomy and support in its dealings with the state.[32] The above might appear to be a rather one-sided version of this relationship—with echoes of the party's infiltration of science witnessed in the Soviet Union—and the political implications of this are examined later. But it was also the case that the rationale behind the Science Pole in Cuba allowed for a rather unexpected working out of this relationship between science and politics. That rationale—the need for cooperation at all levels described by Herrera—in fact contained two rather different demands: (1) integrate the different scientific institutions, which would be necessary to sustain a critical mass of research and production activities (i.e., do what is necessary to foster scientific success), and (2) maintain the original organizational structure of the institution (i.e., do what is necessary to retain political control over the emergent scientific infrastructure). Within this vision therefore was established a simultaneous process of formalization and informalization of scientific conduct and the social organization inscribed within it that would come to be the defining feature of the practice of Cuban biotechnology. It was a tension that rather precisely mirrored that between the nationalist and socialist visions of science embodied in the political rationality of Rectification.

## The Reenchantments of Science

"Might not 'good science,'" sociologist of science Timothy Lenoir asks, "be part of a seamless web of political and economic institutions sustained by sets of value orientations and ideologies?"[33] Many would concur that it might. Such value orientations and ideologies, we might also add, however, are struggled over in the daily politics that accompanies the rhetoric and reality of making spaces for science. Making space for science involves, as we saw with Albert Sasson and Manuel Limonta, epistemic imaginations and orientations as much as architectural or organizational ones, and these are always geographically indexed. Certainly this was evident in the work of Pedro López-Saura and his concern to develop a national version of the International Center for Genetic Engineering and Biotechnology in Cuba, and in Manuel Limonta's vision of a more socially oriented science. In turn, Limonta needed to create a space for experimentation supported (though not controlled) by the Cuban state in order to achieve that vision of a privileged national center set within a broader international space.

In 1986, after several years of such quiet negotiations, the Cuban biotechnology project stepped into the political and scientific limelight with the founding of the CIGB. But it did so at the same time a separate development was also peaking: Rectification—the formation of a particular political rationality as part of Cuba's response to a changing world order, and a reconsideration of where it was situated in that world order. This coming together of biotechnology and Rectification was serendipitous, perhaps, but consequential in two ways. First, as it moved center stage, biotechnology presented a major opportunity for the policy of Rectification to be put into practice. Second, as it did so, the political rationality of Rectification was taken up within the biotechnology project through the promotion of a distinctively nationalist ethos within an otherwise socialist organization. This was implicit, but integral, to the work that would take place within these institutes. The Biological Front's jointly political and scientific work of mobilizing resources, justifying aims, financing projects, supporting results, claiming necessities, and so on made possible the articulation of the political rationality of Rectification within the spaces of the Science Pole.

To cultivate the geographical significance of all this a little more clearly, the various engagements explored within this chapter reveal how the ethical endowments and values of both socialist humanism and nationalist pragmatism were each shaped in different ways but were both distinctly present in the logic that attended the construction of the physical, political, and social space of the Science Pole. In such ways we might say that the physical and social spaces of the Science Pole became mutually constitutive. Once set within this emergent space for biotechnology, however, the relationship between these nationalist and socialist elements began to take on a more specific character. Socialist ideals of science materialized most strongly wherever they also responded to immediate national demands (in the form of the CIGB, for example). This was quite the opposite of the Weberian notion that science is committed to effacement: bound, as it were, to the continual disenchantment that the rationalizing work of science brings. The space for biotechnology that emerged in Cuba in the mid 1980s was one that sought to make possible a very literal reenchantment of science guided by the intersection of pragmatic solutions to idealistic demands.[34] In short, one might respond to Weber by saying that science need not tell us how to live or what to do; the answers we anticipate from it will tell us that. But more crucially still, and as we will explore in the next chapter, the sorts of social and spatial experimentation this brought forth also made possible new ways of thinking about scientific practice itself.

FOUR

# Science City

Gustavo Sierra comes to meet me in the foyer of the Finlay Institute, the CIGB's sister institution, where he is both vice director and head of research. An energetic man in his early fifties, Sierra was heavily involved in the project to develop what is to date the world's only effective meningitis B vaccine. That product is exemplary of the forms of social and scientific organization put into place within the Science Pole toward the end of the 1980s. The Finlay Institute is likewise exemplary of the rapid development of biotechnology in Cuba from the early 1980s to the early 1990s. Like the broader biotechnology project itself, both product and place have widely been feted as an example of what a third-world country can do when an appropriate infrastructure is assembled. But what enabled this infrastructure?

At the end of the previous chapter I suggested that it was not merely the physical infrastructure for doing biotechnology that was put into place in Cuba but a simultaneous approach to science that was formed of different social and political imperatives. Economic and social activity in Cuba is undertaken on the basis of strongly socialized forms of barter and exchange. These are driven by the reality of life on the island: what is described colloquially by various phrases—being in *la lucha*, for example, or simply *sobreviviendo*—and all of which come under the rubric of *sociolismo*: the need to form informal networks of support in order to be able to survive on an island with a rigid, authoritarian government and under the crushing impress of the US embargo. Perhaps the principal feature of *sociolismo* is the reason that almost all Cubans have to resort to *la bolsa negra* (the black market). The relevance of such practices to biotechnology stems from just how widespread they are. It is a widely known fact, for example, that nobody is able to live off the standard state salary, such that all must take part in the black market, in *el bisne*, to some extent. So widely, if informally, acknowledged is this, that such practices become incorporated back into the state itself. Cuban scientific practice needs to be understood in light of these cultural norms. Like all social activity on the island, the social relations of science in the Science Pole cleaved strongly to this improvisational culture.

As the scale of work in the Science Pole expanded, Cuban scientists' knowledge-making practices were given meaning through their ontological commitments to socialism—that, after all, is what Castro had asked for in his speech in 1986. But the informal entrepreneurialism required to meet those commitments sometimes resulted in practices that looked rather more capitalist than socialist.[1] This might, on the surface of it, lead us to question whether Cuban scientists are indeed not best understood as capitalists in disguise, if not to tempt some into the broader belief that biotechnology *is* a necessarily capitalist science. It is more illuminating, however, to ask the opposite: to ask what is really so capitalist about biotech. Of Robert Merton's famous four features of science, one was communalism, meant not in terms of politics but as an ethic of what today we might call communitarianism. As we shall see, the two words are closely linked.[2]

Sierra paints a picture of Cuban biotechnology science after the founding of the CIGB in broad brush. By the late 1980s, the initial work on interferon was continuing apace within the CIGB while, among the many other centers built or refurbished by the Biological Front, the Finlay Institute (opened in 1987) and the Center for Molecular Immunology (CIM, opened in 1991) also came to have particular prominence within the Science Pole. The CIM, the Finlay Institute, and the CIGB have together been described as the "gold standard" of biotechnology in Cuba. Certainly they are the largest, most internationally recognized, and the best funded of the country's biotechnology centers. They are all under the direct administrative control of the Council of State (they are at the heart of politics as much as at the heart of science) and they most clearly embody the mode of operations that I am seeking to characterize. But it was the work to develop a meningitis B vaccine at the Finlay Institute in the mid to late 1980s that is especially characteristic of the practice of biotechnological research that was taking place in the Science Pole at that time. It certainly reveals the nexus between science and politics, on the one hand, and the milieu in which that science was practiced, on the other. But so too does it provide clues as to how and why a particular approach to science developed within Cuba's experimental milieu.

## Foundational Science

With a staff of over nine hundred, the Finlay Institute is comparable in size to the CIGB but has a narrower scientific focus limited to the research and production of human vaccines. Compared to one of Brazil's top biotechnology centers, the Centro de Biotecnología del Instituto Butantan in São Paulo, recently expanded to 700 square meters of installations, the Finlay Institute, with its 20,000 square meters of installations, was built on an impressive scale. The institute was established by the Biological Front to produce and further develop the VA-MENGOC-BC, a combined, recombinant meningitis B and C vaccine. This work, Sierra tells me, "without doubt has been the most successful product [in Cuba] in the last twelve years."[3] Others, including the *Wall Street Journal*, have been even more expansive in their praise of the only proven effective vaccine against meningitis group B in the world. The vaccine even won the World Intellectual Property Organization's (WIPO) gold medal for innovation.[4] As was also typical of the earlier Cuban biotechnology institutions, the Finlay Institute was built entirely around the research and development of just this one drug.

Sierra reminds me of the reasons why the Cubans chose this particular problem to work on. Meningitis, he points out, affects 300,000 people around the world every year; 10 to 14 percent of these cases are fatal, particularly in children of six to twelve months. And what of Cuba within that? From 1916 to 1976 there had been only a few isolated cases in Cuba, but an epidemic broke out in 1976, particularly affecting blood groups B and C. Immunization with an antimeningococcical vaccine from the French pharmaceutical company Mérieux achieved 80 percent coverage and resulted in a substantial decrease in group C cases. But the following year the numbers rose again, this time owing to a resurgence of the group B strain: the strain that, of all the meningitis strains, has proven by far the hardest to develop a vaccine for.

With infection and fatality rates rising, the Ministry of Public Health (MINSAP) set a small group of scientists from the new biotechnology centers to work to find a vaccine against the group B variant. Sierra was part of this group. His comments focus less on the experimental work and more on the clinical trial that sought to demonstrate that the Cubans' candidate vaccine really worked, something which encompassed "a massive double-blind experiment with vaccinated and placebo groups of more than 100,000 students between the age of 10–14 years."[5] Following this, the group tried vaccinating 250,000 young people with the VA-MENGOC-BC vaccine and recorded 95 percent efficacy over all, with 97 percent in the 3 months to

6 years age group.[6] By any standards, this was a substantial level of efficacy. It is possible, however, that in a pharmaceutical market unmediated by the state, a large company would not see fit to then put such a vaccine into the field. They might choose to develop it further first, to ensure that what was finally patented was as advanced as possible; or they may have wanted to know whether the efficacy against a specific local population of patients was replicable in other places.

Because of Cuba's public health model of research, the Finlay Institute was able to proceed with its work on the basis that an immediately available, at least 83 percent effective vaccine proven in the specific Cuban population it would be used in, would save more lives than a more efficacious vaccine (say 95 percent efficacy) which may (or may not) be achievable several years down the line and which could perhaps (or perhaps not) be tested and so engineered to work in other populations too. On the basis of their own 1987–89 clinical trial results, and because of the low reactogenicity of the candidate vaccine, the Cuban Ministry of Public Health thus took the advice of the Scientific Council and proceeded to a mass vaccination campaign with the Finlay Institute's drug in all provinces affected by meningitis B. During 1989–90, over 95 percent of those most at risk were vaccinated—over 3 million people in all. These were mostly children of three to twenty-four months, though vaccinations in people up to twenty years old were carried out. It would later be reported that, after three years, "no severe reactions occurred and one of the most severe epidemics ha[d] been practically eradicated."[7]

The meningitis B project also reveals something of the relationship between the actual *practices* of science and the state, therefore: indeed, this principal success of Cuban biotechnology was in many respects predicated on just that relationship. It was the state (responding immediately to the advice of the State Council) that made possible the speed with which the project moved from research and development through to rollout of the vaccine in Cuba, and these were its winning traits as much as the clinical efficacy of the drug itself. It is interesting to note here that a very similar meningitis B vaccine was being developed and trialed at the same time in Norway. Indeed, both the Norwegians and the Cubans were developing their own versions of an approach to meningitis B (an envelope-based approach) first put forward by American researcher Carl Frasch. Moreover, the Norwegian approach was also a public-health driven approach, and their candidate vaccine was also trialed in schools: the two approaches thus have a good deal in common (and the Cubans and Norwegians had indeed met at various conferences over the previous years). In fact, about the only difference between the two approaches appears to have been that the Norwegians ultimately sought to develop a two-shot vaccine, while the Cubans went for a three-shot vaccine. It is likely this that accounts for the difference in efficacy achieved by the two: Norway's 57 percent not being able to match Cuba's 83 percent. While similar in almost every other respect, therefore, it may just be that cost considerations weighed on the Norwegians' choice of a two-shot vaccine.[8]

As with the interferon group before them, the state sought to capitalize on the work carried out by the meningitis B researchers. They were presented by the media and within scientific circles as modern-day scientific heroes who "dedicated all their energies, talents, and knowledge to the task, with no regard to time or fear." As one would expect, considerable emphasis was given to the personal sacrifice involved: "The human volunteers for the first trials were the researchers themselves, and in a following stage of study on children, their children and those of other researchers of the institute took part."[9] But it was really the extent of the public health system and its deep reach into society that enabled the Cubans to develop not just a functioning theory of how to tackle meningitis B but an achievable and timely delivery of that theory too. These latter elements were as significant an element of the meningitis B vaccination as the laboratory work itself. Such is one defining feature of the Cuban's approach to biotechnology. There were other, less visible, peculiarities developing in the conduct of scientific research in the Science Pole too, however, and they had less to do with the extent to which scientific work might be embedded within the state and more to do with the forms of social and spatial organization that were being put into place there.

## Constitutional Ethics

A clue to some of those peculiarities emerges during a discussion several weeks later with Rolando Pérez, vice director of research at the CIM, which is located on the other side of the Science Pole. Pérez takes me on a tour of his institute and to meet some of the researchers working for him. Later on, in one of the institute's meeting rooms, Pérez will sit down for just a few minutes at a time, before jumping back up to his whiteboard to sketch out an impression of the style of work he sees not just at the CIM but throughout the Science Pole. "Here there is a sort of family atmosphere," he says. "These are very motivated people; it is like an ethic, a raison d'être for them. Social values play a large role for all of us." His colleague, the CIM's Director of Vaccine Research Luis Enrique Fernández, agrees: "I am much more motivated as a basic researcher by the possibility of seeing some results of my work, to be able to see, before I have to retire, that something I have done or worked on has been applied for the betterment of human health."[10] It is as if they are trying to capture something that does not fit into the boxes and arrows of Pérez's flow chart, now complete on the wall. It is also a phrase that reminds me of one of the more pervasive tropes used by scientists who work in the Science Pole to describe the work carried out there, and one that in fact emerged during conversations I had with scientists throughout the Science Pole.

The term *pioneer* has a political meaning in Cuba where it has been introduced into many aspects of the revolution's cultural life: preschoolers are called "Young Pioneers," for example. The term also connects strongly with the mission orientation described above. In discussing biotechnology in a speech in 1992, Castro compared it directly to the revolutionary war in the Sierra Maestra Mountains: "During the war we invented a few things, made certain innovations. It is a miracle that we did not blow ourselves up. . . . In these [revolutionary] times, any results have to be implemented immediately."[11] He was speaking at the National Spare Parts, Equipment, and Advanced Technologies Forum, an important forum in Cuba for keeping outdated machinery going. As much as Castro's comment was intended to be rhetorical, it holds also an important insight that warrants further reflection. The notion of the pioneer has a parallel meaning in the West. In her book *Regional Advantage*, AnnaLee Saxenian talks about a group of technological pioneers thus:

> Drawn together by the challenge of geographic and technological frontiers, the pioneers [of Silicon Valley] created a technical culture that transcended firm and function. They developed less formal social relationships and collaborative traditions that supported experimentation. They created firms that were organized as loosely linked confederations of engineering teams. Without intending to, Silicon Valley's engineers and entrepreneurs were creating a more flexible industrial system, one organized around the region and its professional and technical networks rather than around the individual firm.[12]

This description of Californian entrepreneurs could equally apply to those in the technology milieus of Japan, of France's *Ile de la Cité*, or those in the Third Italy.[13] In all these places authors have drawn attention to "the historically evolved relationship between the internal organization of firms and their connections to one another and to the social structures and institutions of their localities."[14] Science studies scholars have often drawn on parts of this literature in seeking to understand how the forms of social organization that presage scientific innovations are geographically indexed. From such a perspective, sites are not merely points of location, or passive backdrops; they are constitutive elements of phenomena themselves.

Such accounts tend to lean, ultimately, on the role of particular sorts of social norms and values in creative scientific endeavor. On the surface, therefore, there seems little that might reasonably link the kind of socialist scientific practices described in the previous section with the style of flexible production characteristic of advanced capitalist organization in Silicon Valley. But in terms of the sorts of scientific practice and social organization that developed in Cuba there are in fact surprising similarities. Most important, both systems are characterized by the way that particular social values are embedded into their localities. Such constitutional ethics account for the particular association of practice with site, be these what some scholars call "flexible rigidities" or what others call "signatures"—the fact that particular sorts of expertise develop in particular settings.[15] Indeed, just as, at the level of the region, the culture of Silicon Valley promoted an adaptive regional network-based industrial system, and just as, at the level of the firm or laboratory, the technical culture within different laboratories fosters "signatures," so did the technical culture that developed within the Science Pole's space of operations provide the necessary conditions of existence for an alternative modality of scientific conduct. But in Cuba these social values were rather differently constituted and expressed in and through the spaces of the Science Pole.

### Formation

The communities of scientists that developed in the Science Pole during the 1980s were put together rather more in the vein of what Sharon Traweek, in her long-term study of physicists in Japan, has described as a meritocracy: that is, everyone had their place within an overall scheme—all were valued equally for their part.[16] We have already seen in chapter 2 how the Cuban government sends medics abroad as part of an egalitarian internationalist ethos. But the imperative to *realize* material benefits was often also an immediate motivation and would be equally present in the ethical fabric of the community that was taking shape in the science pole. Medicine is conceived within Cuban political discourse as being not only something of social value but of potential economic value too. How did the two ethical imperatives—the socialist humanist and the rather more pragmatic nationalist—play out? And was this enough to replicate the sorts of innovative conditions scholars have observed in high-technology regions in the West?

## Scope and Scale

A certain critical mass of highly trained individuals is often seen as the first prerequisite of any innovation-oriented regional economy. In 1986, the year that the CIGB was opened, there were 39,000 scientific workers in Cuba: one for every 282 people on the island. Around 23,000 of these were involved in research, 16,000 of whom were university graduates.[17] According to Cuban sources from the late 1980s, the average working age of the 4,000 or so of those employed in the institutes of the Science Pole (not all scientists, of course) was then less than twenty-nine.[18] And for an advanced scientific field in a poor country, the gender ratios are very even. At the Finlay Institute, women make up 48.7 percent of its staff of 850.[19] In the United Kingdom, by contrast, only around a third of all jobs on science parks were held by women in the late 1980s.[20] The Cuban experience here in fact compares rather more closely to that of the Soviet Union, where a survey taken in 1978 found that 73 percent of all scientific workers were under forty and over half of them were under thirty-five. Both cases are partly explicable in terms of the desire to channel scientists into large-scale productive efforts, but there are important differences. Those in senior positions in the Soviet Union, for example, were relatively old; over 50 percent of senior researchers with a doctorate were over the age of fifty-five. This was emphatically not the case in Cuba's Science Pole, though one might add that the age of key directors will creep up as they remain in post.[21] Furthermore, while organizational hierarchies were in place in Cuba, the lack of entrenched personnel made it far easier for scientists to work around them. Put simply, there does not appear to have been much room for bureaucrats to be embedded within the scientific infrastructure of the Science Pole, geared as it was to a tight marshaling of resources to immediate national needs, nor was there much time to put them in place administratively as the Science Pole rapidly expanded. In a pattern becoming more familiar, national necessities thereby trumped socialist stringencies in an institutional example of "getting by" when faced with the insuperable structural difficulties—*sociolismo*—that also confront individuals in Cuba.

A sufficient critical mass of scientific personnel existed in the Science Pole, then, but how did these scientists work? It is a question I put to Rolando Pérez: "[These scientists] are mainly twenty-something," he says. "These youngsters can also be in charge of their own projects, their own labs: they have our full confidence. All those who work under me have to get their masters' [degrees] by the age of thirty, and this is the incentive. . . . if they don't want to work here then they move on somewhere else. . . . But as their scientific education is their principal motivation, I can't add to that with salaries or shares." Can that really be? I put the same question some time later to one who has been through this process and recently been promoted to become a director of research. He affirms that researchers are given responsibility at a relatively early stage but elaborates more clearly on the limits to that process.

> Decisions are not made at the laboratory level. The project leaders can decide how to implement it technically, but decisions as to what equipment is needed, how much support is needed and so on are taken at the project level. That happens at my level. We have a committee that meets, and I decide myself sometimes [on its behalf]. Then we have an advisory committee, as well as a business and patents department. These look at the whole picture, feasibility etc, and how many people are needed. After discussion there we then take decisions at the level of Project Management—which is my level. If it is really important then I might go to the director, or the board of directors more generally, which meets every week. For example, we recently decided to reduce our diagnostic profile and increase that on bioinformatics [genomics]. That was decided by us, in research, but with the director. Ideas do however frequently come from researchers, but also from management, from MINSAP even sometimes. Some are suggested by clinicians. As a research director I always try to support that, because if you have doctors on your side then it is much easier to do the clinical trials later on.

## Mobility

A high mobility of labor is also often used as an indicator of the potential for flexible production: ideas circulate with people as they move between jobs and take their experiences with them. Some Cuban scientists have complained of difficulties in leaving to practice their research elsewhere; certainly there is considerable bureaucracy in their way and such movements remain subject to the whim of the state. But many, like José Fernández Britto, president of the Atherosclerosis Society of Cuba, have found opportunities to work abroad.[22] It is considered by many to be one of the perks of the job. Britto, for example, had been to Cambridge, England, four times and had also spent time in Colombia. As a director he also greatly values the system of sending younger researchers to train abroad, as it allows them time out from what is otherwise a demanding schedule of participation in a rapid turnover of research projects. As a result, "obtaining a PhD does not get done so much in Cuba—those that have them are more likely to have done them abroad," another researcher, Morales, told me. I have since examined bibliographical data on the careers of sixty-four scientists in Cuba and the results did not support this. The data set I examined suggests that 34 percent of scientists do their PhDs abroad.[23] It does, however, confirm the perhaps more important point—that a large number of scientists have actively worked abroad, on average in two countries from this sample. Given their young average age, this is an impressive figure for a country such as Cuba. Similarly, attendance at international scientific conferences has long been valued. While Cuban scientists are therefore able to benefit from training abroad, the brain drain that so often accompanies such movements and that is so costly for poorer countries, particularly in the fields of science and technology, is not such a problem in Cuba.[24] This has been an essential part of the reason Cuba was able to develop such a concentration of highly qualified personnel.

## A Biomedical Emphasis

Pérez's own experience of doing research in Cuba is typical, and as a director he has a say in how research structures operate.

> I began as a physicist, in the 1970s, but then Agustín Lage [the director of the CIM] encouraged me to switch to biology, which was then a very exciting area to work in. I undertook a transitional thesis in 1973–74 at CENIC, where I took the courses I needed on biology and biochemistry. I took part in the courses they were providing for medics, to improve their basic science. I then took my master's in molecular biology, and then my doctorate, which I did in Nice, in France, on the cancerization of cells. I defended my thesis in Cuba in 1982 and moved into work at INOR [Cuba's National Institute for Oncology and Radiology] and then the CIM.

Other scientists I spoke to had similar experiences of switching from one specialty to another as both need and possibility arose. One even noted that in his university days many of his fellow students were invited to make transitions; in his own case, it was due to a glut of biomedical scientists and an insufficient number of veterinary scientists. As a result of these transfers, figures showing merely who studied what are not as revealing as they might be: many scientists now classified as physicists or chemists first trained as biologists. And one might speculate that such personal interests formed in biology, and in conjunction with a research environment that was highly receptive to biology, helped foster productive cross-disciplinary links between researchers who understood each other's subjects well. Biology has certainly been a sort of common language among specialists, helping to overcome the barriers of excessive specialization.[25] "The solidarity of specialist communities—or such solidarity as exists—is coordinated through their specialist knowledge," as Steven Shapin reminds us.[26] But so too is it coordinated through the same social structures by which they can know that they are "special." These structures are found throughout society and not just in the lab, of course. In Cuba, this is apparent. What is happening in the Science Pole is, in some respects, a microcosm of what is happening in Cuba more broadly. "For there to be solutions to the problem of knowledge," Shapin elaborates, "there have to be practical solutions to the problems of trust, authority, and moral order."[27] This points to a broader issue where the mobilization of a particular sort of work ethic becomes important: the formation not just of a technically skilled and competent workforce but of a trusting, or at least morally ordered workforce too. How, then, was this crafted in Cuba?

## Conduct

Between research visits to Cuba I spoke to a former CIGB scientist who now lives and works in the United States, having left because of his dissatisfaction with the more overt politicization of work practices discussed in chapter 7. Jorge was a younger worker—just the sort of figure Rolando Pérez was commenting on. Interviews for access to the CIGB, he tells me, are held over a period of several weeks (up to three months) and candidates must pass monthly evaluations for six months before becoming permanent staff. But assessment does not stop there. My conversation with Jorge then turned to the role of the individual in the way research was organized:

SRH: So what sort of group were you were working in?

JP: It was mainly my boss, I, and two other researchers.

SRH: And were there many other groups of that size that were working? Was that a fairly standard size of research group that people in the institute would work in?

JP: Yes I think so, maybe four or five people.

SRH: And these small groups . . . who would they report the findings to when they came up with them?

JP: They have, like, every year they have to give a seminar to the division they were in. . . . At some point they also [began to] require the scientist to have at least one publication per year, so people have to work on producing their own papers.

SRH: So in this sense, the scientists are being evaluated specifically according to their output?
JP: Yes, they receive evaluation yeah, I don't remember exactly the grading, but it would be like "Good, OK, or Bad."

The procedure this scientist describes presents a picture with which scientists in many other countries are familiar: publish or perish. But in the CIGB a particularly high premium was placed on results. The mission orientation was something engendered at every stage of a scientist's career. It was also something given legislative support from a 1987 agreement of the Council of Ministers entitled "Regulation of Discipline and Selection of Personnel in Scientific Research Units."[28]

### *Consagración*: Conduct of the Self

We might say that in Cuba such conditioning of labor was organized around a particular form of ethical reflexivity, what Cuban scientists refer to as *consagración*. It certainly seems to be via just such a form of conduct that the ideal of socialist humanism is embodied. The existence of such an ideal does not mean that every scientist adhered to it, of course, but its codes helped set the framework in which they worked. Subjectivity matters in science because what we know and understand about others is very much a part of what we know and understand about things: the "people world" of society and the "thing world" of science are not separate. It was a point Gustavo Sierra emphasized during my visit to the Finlay Institute. During a conversation about work on the meningitis B vaccine, we got on to the topic of incentives in Cuban science, and Sierra offered a clear articulation of what *consagración* is all about. "People here work not for material rewards but to help their country . . . not all want to, and those that don't leave, attracted by the incentives elsewhere, but, well . . . the majority don't work for money."[29]

Weber would no doubt approve. In the Cuban workplace the word *consagración* is used to describe (and by naming it, one suspects, enforce) "painstaking and total dedication to work."[30] It is in effect a reworking of the New Man—the "moral man" postulated by Che Guevara in the 1960s—as a sort of Good Scientist. More than a few of Guevara's critics have pointed to his dogmatism. But in the Soviet Union, the party's attempts to promote communist thinking in scientific circles involved more extreme processes of conversion and co-option, an approach most famously embodied in the Lysenko affair.[31] The scientists were either converted into believing in the official line on science, or they were co-opted into adhering to it.[32] In Cuba, however, the state's attempt to promote socialist principles in workplace practice and in the domain of ethics revolves around the more inculcative emphasis on self-conduct. This has echoes of former religious modes of conditioning labor (Catholic consecration or more puritan forms of work ethic, for example) and is far from unproblematic itself, of course. Few of the scientists I regularly spent time with, and spoke to about their jobs, worked for such an abstract notion of their country—though some certainly did work with a very strong idea of the presumed social benefits of their research. On the whole they worked because they enjoyed the work itself, the satisfaction it gave them, and the opportunities that they found through it and in science more generally.[33] As already mentioned, many had been able to travel or study abroad, or they were provided with accommodation. Many were also given a cherished independence *within* research groups to pursue their own intellectual development. There are thus both practical and principled factors at play.

Commenting on just this aspect of scientific protocol, a review of Cuban research in the journal *Science and Policy* concluded, with respect to Cuban oceanographers, that they displayed "more bottom up initiative than is normally expected in a Marxist-Leninist state."[34] This is not to deny the presence of hierarchies, but it does suggest that they do not always stifle creativity. As one biomedical scientist related to me, for example, when there were equipment shortages in the early 1990s, she resorted to making her own containers for conducting a form of staining experiment. She made them with the help of her husband, who was a physicist, using a blowtorch and some degraded oil to shape the container in the required way. The idea caught on and she ended up making a whole set for colleagues too. She joked that she might have made a profit. She also commented that "you do have shortages and problems and difficulties always and it stimulates creativity, there's no doubt about that. There are many more examples of things in the laboratory that we use creatively." Again, this sort of account has been noted in the Soviet experience, but the process of informalization—by which I mean, quite literally, getting by and getting on, by inventive means if need be—has been significantly stronger in Cuba. And it does start to look similar to the work being undertaken within the new industrial regions of the capitalist West.[35] The reason why hinges on the formation of particular social networks and the ways they are put into place.

## *Laboratory Lifeworlds*

A sense of community can play an important role in scientific and technological innovation: it helps to foster commonly accepted principles, aims, and methods. Not surprisingly, perhaps, many of the scientists I spoke to made frequent mention of Cuba's "human capital." In Putnam's classic study, the central premise of human or social capital is that social networks have value. Social capital refers to the collective value of all "social networks" (who people know) and the inclinations that arise from these networks to do things for each other (norms of reciprocity).[36] In this regard, Carlos Mella of the CIGB is perhaps right to argue that "socialism *does* have certain [economic] advantages compared to other social systems." But to have a purchase on the world, social relationships need to be solidified through institutional, economic, political, or other regulative means. That is to say, at some level they must obtain material expression. In a classic essay, the economic sociologist Mark Granovetter uses the term *embedded* to refer to just this sort of materialization of social lifeworlds within particular geographical contexts.[37] It is a pertinent insight, but the grain of the world also catches back against the social relationships which give it form: communities develop in dialogue with the world, not just other members of their immediate community. This took a rather particular form in the Science Pole.

At the opening of the CIGB, it had been announced that accommodation would also be provided with the centers. Since then, between two and three thousand units have been constructed between the CIGB and the Frank País Hospital toward the southwest of Havana "such that, in a certain way," it was announced, "a scientific community is going to be created here: you [the scientists] will live so close that you only need walk to work."[38] This "community building," however, also serves a different end: it contributes directly to the formal organization and administration of work in the Science Pole, and thereby also to the structuring of informal responses. It also creates real communities of individuals who trust one another and thereby "make better cooperative relations," to paraphrase Gambetta.[39] It embeds the scientists more firmly within the milieu of the Science Pole and ties them more closely into the political rationality that circumscribes it. The scientists were located proximate to, if not right next door to, the laboratories in which they worked. For those who live in other parts of the city, special buses take people to and from work. Facilities such as daycare for children were also provided, as they would be at many research-based workplaces, such as university campuses, in the West. With labor localized in this way, the Scientific Council (which by the late 1980s had become the operational body of the Biological Front) then set about determining research objectives on the basis of fourteen-hour daily work schedules, as standard.[40] But it is not just laboratory work that these schedules tie the scientists to; as a result they also spend more time in their own tightly knit communities—talking about work and sharing ideas. These are precisely the same, rather informal elements at the heart of other successful regional cultures of innovation, such as Silicon Valley.

But is all of this simply a case of Stokanovich redux in Cuba? While it is true, for example, that none of the scientists were obliged to locate in the housing built for them, given the young age of many, the shortage of housing in Havana, and the difficulties of moving, it is likely that if they did not live at the Science Pole, they would be living with their parents. Most scientists I visited who were not living in the Science Pole were indeed living with their parents (unless they were of an older generation). Not only was it in some senses easier to live in the Science Pole, it was also often more desirable and provided another "life opportunity" in a country where, primarily thanks to the embargo, such possibilities can be rare. Certainly that was the case for Maria, a young researcher who worked at the CIGB on bioinformatics. Shortly before my arrival she had moved into the accommodation block opposite the CIGB with her young son, having shared with her mother until that point.

All this also has important implications for the formation of a culture of risk-taking: another important component of high-technology research. It is particularly important in biotechnology, where start-ups more usually fail than succeed. Here the situation in Cuba very closely mirrors that in Silicon Valley, where "the region's culture encouraged risk and accepted failure."[41] In Cuba, risk taking is often "glorified."[42] To return to our analogy, the operationalization of risk in Silicon Valley and Cuba's Science Pole is somewhat different. In Silicon Valley, risks were taken in betting on a product, or a company, and chancing financial ruin if the venture failed. It was the regional culture that ensured that the failure of a particular venture did not necessarily mean financial ruin for the individual. As an *individual*, you might be able to get another job the very next week. When this sort of activity is scaled up to the social level, the result is a high turnover of employees, which in turn means a rapid circulation of information, which in turn promotes institutional flexibility and adaptability. In the Science Pole, however, it is the *risk*, not the individual, that is displaced. Individual scientists would be unlikely to lose their jobs, because they were working for the state. The formation of scientific labor, the means through which this human material was then regulated, and the ethic of improvisation that resulted made possible within the space of the Science Pole a risk culture that might otherwise have been less prominent in a socialist context. But as Paul Rabinow notes in trying to get a handle on the peculiar qualities of the scientific entrepreneur, "It is not mere business [or scientific or managerial] astuteness, that sort of thing is common enough; it is an ethos . . . an ethically colored maxim for the conduct of life. This is the quality that interests us."[43] Was such an ethos present in Cuba's Science Pole?

## A Form of Experiment

In her office, many thousands of miles away, working in a rather different culture of soft-money science in the UK, Miriam sheds light on the operation of a distinctively Cuban ethos of research and, furthermore, points to how such an ethos may be bound into a particular cultural framework:

> I think it has a lot to do with what you might call "connection." Through this connection we all learn from other experiences and it is a classical situation which always came to mind as an adolescent and a young scientist telling myself: we must be able to achieve whatever we work hard for, bringing to mind the passage of the twelve men left after the first defeat of the Cuban rebels when Fidel said to Raúl, "we are enough to win the revolution." You don't give up, just work harder. And I think that the confidence of the Cuban scientists always behind any particular project had the root in that conviction. This conviction is a mind power.

Conditions of necessity can be a driver for improvisation and sometimes innovation also. Something similar happened with the forms of social relationship that emerged in the Science Pole. They did so because of the way that the sort of ontological commitments to socialism that Miriam's comments allude to were articulated (or enacted, we might say) within the spaces of the Science Pole. There are any number of ways in which such commitments were enacted, of course, but we might consider as critical among them the formation of scientific networks, the operation of centers of calculation, the development of basic research within an applied framework, and the epistemic effects of social and political boundaries.

### *Networks: Integration*

Ernesto Bravo is a professor at Havana's College of Medical Sciences. An Argentinian by birth, he has worked in Cuba for many years and has taken a leading role in biomedical and health-care services. He is a strong believer in the work being done in Cuba. In 1983 he founded the North American–

Cuban Scientific Exchange Program (NACSEX). As a formal organization, its purpose is to foster scientific exchange between two countries otherwise locked into a suffocating embrace. Clearly this is one area where being socialist and subject to an embargo by the United States offers no regional advantage whatsoever, which is what makes such points of transfer as Bravo's exchange program so important. It offers a means of in-sourcing the sort of knowledge that can be hard to come by in Cuba: industry gossip, the latest rumors as to what's hot, and so on. To date NACSEX has managed to bring more than eighty first-rate US scientists to Cuba, usually on an individual basis, though it has managed to secure just one visa for a Cuban scientist to visit the United States. Despite the fact that NACSEX is sponsored by the Cuban Academy of Sciences (ACC), it has maintained a certain space of operations independent of government policy.[44] But as many countries have found out the hard way, access to information is not enough. That information itself needs to be put to work. Here, the much vaunted cooperation between the various centers that began toward the end of the 1980s was another crucial element in Cuba's experimental milieu.[45] This, after all, is the whole point of the Science Pole.

Miriam's experiences are indicative of this. When one of the original scientists in the interferon project, and at that time a key figure at the Science Pole, asked her in 1988 whether she would be interested in heading up a new Center for Pharmaceutical Chemistry, she replied,

> I told him yes, but that I was a pharmacologist and the problem was that we didn't have the conditions to validate any of the products. His proposition was why not share the facilities for doing preclinical studies that they had at CIGB. Selman was part of the Grupo de Apoyo. This is something you don't really have elsewhere. The group consists mainly of politicians, but politicians who understand science of course. They are the most antibureaucratic people ever. Whatever you want, they can get for you. In the case of the proposed Center for Pharmaceutical Chemistry, everything was arranged within hours for me and my group to carry out all the preclinical work we would need to do at CIGB. So we did that. We did preclinical work on gamma interferon, and on streptokinase for them, and we got to use their facilities in return. They also made sure that if you needed to recruit someone, they would find the best people to work with. They might be untrained, or not formally educated in the particular area you wanted, and they might be people from Hungary, or Czechoslovakia or wherever, but once they were recruited they would hit the ground running. They would work from 8:00 a.m. to 10:00 p.m. and you would mold them into the scientist they could be.

There can be many damaging features of selective cabals such as the Grupo de Apoyo, and one suspects it might just as easily end a particular project as start one up. But its work here represents just one example of the broader social connectivity of the Science Pole that I am trying to elucidate. It is not unusual, of course, that scientists should cooperate willingly, as Miriam points to here. What is unusual is the extent to which this takes place, and that is in part a function of the sense of community that exists in the Science Pole. "It is absolutely impossible," says Pedro López-Saura, "for you to find a project that is worked on [here] in just one institute." Again, the intention here is not to overlook the more problematic aspects of these same processes. Institutes in the Science Pole often scrap over who gets the best students, the latest technology, and the scarcest resources, for example. But the more general way in which they are *also* able to cooperate intensively was what worked in Silicon Valley, and, albeit in a rather different way, it appears also to be what works in Cuba.[46]

## *Centers of Calculation*

The body responsible for organizing these various networks of scientists, machinery, protocol, and product—the rather reclusive Scientific Council—adds a further, complicating dimension to all this. The council meets around once a month, attended by the directors of each of the institutes of the Science Pole. The idea is to make sure that everyone knows who is doing what and why, and who needs what and when. The work of the Scientific Council was often going on in the background of my own research. As I crisscrossed from one institute to the next, quite often I would pass the directors of the various institutes as they were ferried from one meeting to another in state Ladas. If the comments of several scientists are anything to go by, this is not a place for sipping coffee and making a to-do list. Decisions are rapid-fire and actions are expected.

Take one typical Monday morning, for example. It was not yet 8:15, and the entrance gate to the CIGB, where I was waiting to pick up my entry pass for a meeting with López-Saura, was already awash with activity. There was the usual flow of casually dressed scientists into the complex. Hardly anyone wears a suit at the CIGB. Luis Herrera was on this day, as was some of his immediate entourage: López-Saura, Sonia Negrín, Borroto, and the other vice directors who inhabit the ground floor corridor of the north wing. But almost everyone else was in short sleeves and cotton slacks. I'd been warned that López-Saura would be busy: it was the first Monday of the month when the Scientific Council, of which López-Saura is a member, holds its meeting. And before we met at 4:00 in the afternoon—I had arrived at 8:00 in the morning to see him—I would see López-Saura drive past three or four times, squashed into a Lada with *Estado* [State Council] stenciled onto the side. I learned later that the Scientific Council had been discussing proposals for further integrating the work at CIGB with the work at the Finlay Institute. The Scientific Council discusses all the work throughout the Science Pole and makes proposals as to what should be prioritized or changed. There can't be more than fifteen or so on the council. When we do finally meet, López-Saura describes for me the general outlines of what they have been doing. He does not go into detail:

> There are generally two ways that a new project might emerge: it might be deemed necessary, or someone may come up with an idea or a solution for a problem that already exists. . . . This is what we then discuss on the Scientific Council, which considers projects from throughout the Science Pole. It conducts its work via commissions.

A few days later at the Finlay Institute this use of generalities seems to be repeated. One of Sierra's assistants describes to me how he has been discussing of late a project over at CIGB. I am not told what that project might be. There is also to be a conference later on this week. That's public information.

The sorts of commissions López-Saura mentions often involve fact-finding missions by junior scientists, who are invited to make suggestions as to how best to go about things. Once the policy is decided and the ends set, then the appropriate means are up for debate. Here, then, is the center of calculation. Here is where the determination of determinations takes place.

One of the thornier determinations to be made at the Scientific Council is the issue of funding. Cuba is not a rich country and, while biotechnology may be prioritized, money remains tight (pressingly so, as we will see in later chapters). As a result, the allocation of funding between centers, and at difficult times one presumes from one center to another, works against the trend of cooperation to in fact differentiate between the various institutes of the Science Pole. Funding is a notoriously difficult area about which to gain reliable information in socialist Cuba, but at least something can be said about the context in which this particular determination was made.

In 1985 just prior to the emergence of Rectification, it was decided to introduce monetary relations to scientific work, effectively providing for limited autonomy and a controlled incentive structure.[47] Hence, Resolution 2/85 was passed in 1985 precisely with the aim of fostering a limited competitive environment: "In factories which achieve economically favorable results through the application of innovations or rationalizations, a fund to the value of 20 percent of the economic gain will be created" and that returned to the factory in question.[48] Again, it is a formula that contributes toward replicating in socialist Cuba something of the sort of productively competitive structure that has been identified as an important feature of many high-technology sectors in the West. But at the same time as these centers were being both integrated and encouraged to compete, they were also being kept apart from the often centralizing requirements made of the rest of the productive sector.[49] The crucial element in all of this is the distinctiveness of the Science Pole as a specific arena of activity. The Science Pole provided not simply a physical space for biotechnology research—a clustering of research institutes sharing resources—but also a reconfigured form of political space in which normal procedures of socialism were locally reworked. In short, some forms of scientific conduct are taken on and others are resisted; this is achieved through a prior organization and operationalization of sociospatial arrangements.

### *Applied versus Basic Research*

Applied research is one of the hallmarks of the Cuban development of biotechnology. Carlos Castañeda, director at the gastroenterology unit and actively involved in research at the CIGB, is a strong advocate of the importance of applied research and the advantages he sees that this has had for Cuban biomedicine. But how does this square with the need to carry out the basic research necessary for successful innovation? During a conversation about the research then taking place at the CIM, Rolando Pérez offered a similar account but added some illuminating details:

> In the first place, when we create new institutions, they are always created around experienced groups [*colectivos*]. So a bridge is made from one center to the next. The core in any center is an established group from elsewhere, but the surrounding work is new, a sort of spin-off. This is how new experimental groups develop. Second, with respect to the strategy we adopt, while the developed world countries have a "push" model, from academic research to industry, here the strategy is to have a "pull" model, science pulls aca-

demic, or basic research along with it, and pulls industry along with it. I believe this is a relatively unique experience and it stems from these research groups.

The figures bear the thrust of this out: in 1980, for example, while there were only 166 scientists working on strictly basic research, there were 2,820 developing basic research as part of "integrated," that is to say, applied, projects.[50] So too might we consider that sending scientists abroad is also a spatial fix—in the way that David Harvey uses the term—to the problem of combining basic with applied research in the absence of extensive funding. While the framework for research is predominantly applied in Cuba (83 percent by one figure), the researchers have in many cases done basic research, but abroad. Something not dissimilar and on a larger scale is currently afoot with the return of skilled professionals from Western countries to India.

But Pérez's answer also turns on definitions of basic and applied research. As set out in the introduction, basic research is usually seen as primarily concerned with innovation and producing new "knowledge," while applied research is often seen as a more technical enterprise, applying already tested knowledge to practical effect. A report published earlier in the decade gave a fairly blunt interpretation of the relationship between basic and applied research in Latin American biomedical research. The relationship was characterized, it said, by "a situation in which the geneticist accused the politician of a lack of interest in scientific development, while the politician accused the scientist of dedicating himself to activities far-removed from the 'pressing' needs of the country."[51] Pure genetic research, it was therefore imagined, would likely stay within the university while outside it geneticists might find room to develop more applied approaches. It pointed to a coming golden era in which the pursuit of applied research would ultimately lead to specialized areas of basic research. A case of "grub first, then ethics" perhaps, as Bertolt Brecht put it. But as a diagnosis of the situation and an aspiration, it captures only some of what happened in Cuba.

In Cuba, as in much of Latin America it seems, basic science meets ontological resistance. This is not to say that it is rejected outright. Rather, it appears to be perpetually negotiated and reformulated within a concept of applied science. This indeed is the hallmark of the particular scientific rationality that developed in the Science Pole at the intersection of socialist humanism and nationalist pragmatism. This is the nub of what it was that marked *as unusual* the research on meningitis B, or on interferon, or any of the other projects taking place in the Science Pole. "The project of cognition," as Herbert Marcuse pointed out, "involves operations on objects, or abstractions from objects which occur in a universe of discourse and action. Science observes, calculates and theorizes from a position in this universe."[52] The CIM's experience with monoclonal antibodies only reinforces this point: while their applied research was for long considered mere technological tinkering, by the late 1990s the degree to which they could manipulate immunoregulatory responses as a direct result of a decade of this tinkering put them, as we shall see in the following chapter, at a new cutting edge in cancer research. Such paradigm shifts are essential to scientific development. The epistemic reconfigurations they rely on, however, stem from the development and solidification of new forms of collective scientific subjectivity. And it was precisely this that could be seen in the approach to basic versus applied research in Cuba.

### *Boundaries*

I have described above some rather explicitly sociological elements that together might be taken as indicative of a distinctively Cuban practice. I have also suggested that what is crucial in all of this is the way that each was shaped by the promotion of a particular sort of ethic, on the one hand, and the ways that the social and spatial organization of the Science Pole shaped the realization of that ethic at the level of scientific practice, on the other. It warrants describing the latter in more detail, for nowhere is the constitutive role of space in scientific endeavor clearer than in the way that the Science Pole delimited a particular sort of epistemic space. This is not merely coincidental with the fact that scientific practice took on a particular mode of operation within it. The Science Pole was itself a sort of delimited space, set into a broader socialist polity, yet unbundled from some of its strictures and plugged into wider global scientific currents. It was a site of epistemic tension, and it was scientifically febrile as a result. Such boundary effects made the science pole itself into an experimental form.

## An Experimental Form

What then does it mean to produce a particular sort of knowledge—in this case, pharmaceutical knowledge—as a "moral man" acting on an openly crafted plane of social activity? It meant, as I have tried to set out, novel forms of citizenship and social engagement in the Science Pole at the same time as it meant a particularly ethically flavored conduct with respect to scientific practice. This had important implications in terms of the forms of practical epistemology deployed within the Science Pole. There, outward commitment to socialism, for example, might officially be read as a guarantee of credibility to speak for or about science. But, in fact, rather more emphasis appears to have been placed on practical experience. For this reason, hierarchies were often dissipated. If, as we will see, Cuban science was later presented to the world as a secret to be discovered, this was one part of the secret that remains slow to come to light. The result, for a while, was a highly efficacious scientific milieu.

The extent to which Cuban bioscience in the 1980s was able to develop a capacity for innovation, despite many of the usual difficulties faced by a poorer nation, appears to have been underpinned by important forms of socialization, some of which I have tried to outline here. These forms of socialization were not of the sort that Thomas Kuhn identified, however—that is to say, of regulative principles of social order being determined on the basis of particular model, or exemplary, forms of scientific methods (such as the production of DNA, for example). Rather, the socializing influence of these centers was focused on producing a particular sort of scientist—a morally bound and socially maximized individual, namely, the Good Scientist we met earlier. This closely echoed the way that the "moral man" debates and Rectification sought to work on the human material of the population, on human subjectivity first, in the hope that this would produce a particular sort of state, rather than the other way round (as is usual in authoritarian states). We are not talking then about forms of regulation applied to foster particular forms of conduct, but about particular forms of conduct that emerge when people use local frames of meaning to interpret and forge their understanding of particular things (like scientific theories): I met many good scientists in Cuba, but none were as socially flat as the ideal of the Good Scientist. The context in which one works thus becomes integral to what one is doing.

We can be rather more specific about these findings with respect to Cuba. The realization of scientific work within the Science Pole in the late 1980s (and into the 1990s) witnessed the binding of a particular ethic to a particular place; it set that structure within a broader machinery of state that also encompassed structures of care, governmental institutions, and the like; and it posited the two (that is, the state and science) in a sort of productive tension that could only be held in place within the physical bounds of the Science Pole. In this sense, the Science Pole was indeed a form of science city—an epistemic and social milieu within which biotechnology took place as both a practice and a culture, and as such a spatial form it set the literal bounds for the forms of scientific activity that were taking place within it. It therefore needs to be understood as more than a simple transposition of the model of regional innovation pioneered in Silicon Valley: there is an economic geography at work here but also an epistemic geography. For one, there are numerous obvious examples where Cuban biotechnology's experimental milieu conforms more strongly to less flexible models of industrial organization: the existence of vertical hierarchies, for example, or the encouragement of stability and self-reliance, and the internalization of a wide range of productive activities within larger institutions. For another, that was simply never the plan. There is no documented intention of copying Western models of innovation promotion; the emergence of the Science Pole and the development of particular approaches to science within it was an autochthonous affair. How then are we to understand this particular space for experiment?

For Pérez, Cuban biotechnology is best understood as located within a particular "space of creativity and innovation," a nonliteral space that makes possible "a kind of research at the frontier between basic and applied science." Under this rubric, while there may have been relatively few outlets for any burgeoning entrepreneurial culture in Cuba, the principal effect of such a culture, namely the promotion of risk-taking, was achieved through alternative means (such as the internalization of risk by the state). Within the Science Pole, processes of integration could be productively set alongside processes of differentiation, providing for a productive tension between, for example, centers that were competing for funding yet sharing results, or between formal vertical hierarchies and informal horizontal collaboration.

The result was a highly innovative conjuncture in which Cuban biotechnology came to develop just the pragmatic mode of science John Dewey believed in.[53] This meant that the differently framed ends laid down as the objectives of research and development (that is, national needs to be met through socialist channels) allowed for experimentation in the realm of ethics that supports them (the development of efficient but unusual production methods, for example, or the sharing of resources between centers and so on). For reasons of necessity (such as limited finance) and contingency (such as the desire to apply the results for political ends), Cuban scientists put into place, during the 1980s at least, a very experimental milieu: one whose *informality* required the very formality of the social and spatial organization of the Science Pole, whose means were flexible (one could approach cancer research in terms of a therapeutic model even if the global trend at the time was to conceive of it as a material problem requiring various ways to "cut it, burn it, kill it," as the saying goes), but whose ends were not (if a treatment was required in three years, then some sort of treatment, even if not the most efficacious or marketable, would have to be found—as with the work on meningitis B).

But there is a catch here that ties these developments back to the geography in which they are set. Alone, neither these informal elements nor the formal ones could have supported the sort of innovative milieu I am describing. We cannot understand these developments by reference to the laboratory alone, for it was the broader context of the Science Pole that provided the space for both sets of processes to coalesce around the formation of a particular sort of scientist. The Science Pole was not merely an elaborate network of sites of scientific inscription therefore. More properly speaking, its space of operations was produced by the struggle, characteristic of this rationality, between official ideology and the scientists who sought to work around that ideology in their practices in order to fulfill their own objectives as they perpetuated the system which made those objectives possible or even thinkable. The Science Pole was a particular sort of epistemic space whose boundaries were just as porous as the physical site, but also just as real. It was the achievement of this particular space for experiment that lay at the heart of Cuba's remarkable early successes with biotechnology, including the celebrated production of a meningitis B vaccine. Quite simply, by the late 1980s, Castro's vision that the future of the nation had necessarily to be a future of men of science had been realized in a form that no one—and perhaps least of all himself—would have predicted.

And by the end of the decade it wasn't just the Cubans who were taking note of what was now being more commonly referred to as a distinctively Cuban biotechnology. There was in fact considerable agreement among both Cuban and foreign scientists that this was something of a success story, a paradigm of autonomous biotechnology conceived as an alternative to the dominant model of privatized, large-scale research that typifies the approach of Western firms like Monsanto. Even the politicians agreed. At Cuba's preeminent business fair in 1989, the mood was upbeat. "It is possible to predict," it was claimed, "that the biotechnology industry will make a significant contribution to the economic and social development of the country over the next half decade."[54] Meanwhile, across Latin America, countries looked to Cuba as a model of what could be achieved. Biotechnology expert Daniel Goldstein noted that "all of Latin America together cannot reach the ankle of the Cuban endeavor" in biotechnology, and it was already possible to see how Cuban biotechnology would come to be described more generally as the "backbone" of the island's so-called entrepreneurial socialism.[55] But if all this served as a reminder that alternative economic (and scientific) models can prosper at the margins of US-centered global neoliberalism, Cuba's luxury of distance was not to be enjoyed for long.

FIVE

# Sticky History

In a broadcast on ABC, some ten years after the event, Agustín Lage, director of the Center for Molecular Immunology in Havana, spoke candidly of the problems Cuba had confronted in the early 1990s. "To notice that we have economic difficulties in Cuba is not a discovery," Lage observed. "Everybody knows the reason for that. This economic war we are suffering from has an impact in every sphere of life."[1] Lage wanted most of all to refer to the US embargo. As brother of then Vice President Carlos Lage, one would expect nothing less. But his comments were also directed at the economic crisis unleashed in Cuba by the fall of the Soviet bloc. The collapse of Soviet socialism was a catastrophic event for the island. In 1989 Cuba conducted as much as 83.1 percent of its trade within the Council for Mutual Economic Assistance (CMEA), the socialist trading bloc.[2] Then, "almost from a Monday to a Tuesday," as Carlos Borroto of the CIGB says, the Soviet Union collapsed.[3] As if overnight, Castro lost most of his credit, half of his oil deliveries, and 80 percent of his export markets. On the streets, cars ran out of gas and good meat became nearly impossible to obtain.

By 1992 national income had dropped by around 40 percent, and in 1993 students took to the streets in protest until Castro himself was finally obliged to intervene.[4] A response to the escalating crisis was sorely needed. It came with the announcement of an emergency program known as the "Special Period in Time of Peace." Much has been written on the Special Period. Some describe it as nothing more than a trick of nomenclature devised by Castro to displace the real reasons for economic crisis onto external economic and political conditions. Others see it as a new stage in the development of the revolution. For our purposes, however, it is perhaps best considered alongside Rectification as an emergent form of political rationality: a series of policies and technologies of rule focused on securing the immediate survival of the regime and safeguarding the principal achievements of the revolution.

The new political rationality was officially pronounced at the Fourth Party Congress, in 1994.[5] It ushered in a series of political and economic reforms intended to mitigate the worst of the economic crisis. The rationing of everything from food to electricity was the most basic of these policies, but they included more subtle and longer-term measures bearing all the hallmarks of Cuban socialism: urban gardening initiatives, car-share schemes, and the like. At the same time there was a limited degree of economic liberalization.[6] State ownership of Cuban entities was to remain but limited joint ventures were encouraged. The effects of this emergent rationality did not simply unfold on the streets or in the offices at JUCEPLAN, however. They unfolded in the scientific realm too and with profound consequence. For if it had not exactly been isolated from broader developments in global science, Cuban science had until now been at least introspective. Targets were set on the basis of specifically local needs and, so far as we might summarize the Cubans' approach to biotechnology, these ends were privileged over the means. All this was part of the reason the Cubans developed what I have described as a particular modality of biotechnology, and that modality in turn has a lot to say about why the Cubans were able to achieve quite substantial scientific breakthroughs without the inputs of capital and the vast resources of labor and materials available to similarly sized biotech efforts in the West.

But the fall of the Soviet bloc brought Cuba's formerly secluded biotechnology into contact with all the heavily capitalized features of Western science—strict intellectual property rights, competitive frameworks, shareholder commitments, and so on—at the same time as it brought the country politically, and economically, to its knees. Biotechnology, alongside tourism and sugar, in fact became one of three industries selected to embody the new reforms.[7] Specifically, it was hoped that biotechnology, particularly its agricultural applications, might support the new National Food Program and even provide significant exports. In various ways, this new economic imperative insinuated itself into the realm of scientific practice: diplomas for "contribution to the economic development of the country" awarded to some of the institutes, for example. But against such demands, the priority for the scientists and bureaucrats in charge of the Science Pole was simply to try and maintain a space for their science. What surfaced in the early 1990s, therefore, was a multifaceted struggle between Cuba's locally nurtured epistemic milieu and the machinery of the global pharmaceutical economy mediated through the circumstances of a localized political crisis.

## Capitalizing Science

In a quietly located corner of the Science Pole, there lies an institution whose own emergence effectively charts this drive for limited commercialization. Biomundi was the first Cuban life-sciences consulting organization. Founded in 1992, it came to play an important role in the reorientation of Cuban biotechnology toward international norms, providing information to the institutes of the Science Pole on a wide range of countries and organizations involved in biotech. At the same time it set about developing a public perception of biotechnology in Cuba, particularly through the state organs of *Granma*, *Juventud Rebelde*, and *Trabajadores*. Through the work of Biomundi, the yearly biotechnology conferences, and the well-attended annual Havana trade fairs, biotechnology did indeed become a visible part of Cuba's interface with Western investors and a central pillar of Castro's strategy for dealing with external sector imbalances.[8] But all this did precious little to obscure the fact that these were lean years for Cuban science. In just the most visible example, construction of a nuclear power station in Camagüey was mothballed, while investment in scientific infrastructure and equipment also declined dramatically.[9] The generalized economic crisis may not have dampened the government's hopes for biotechnology, but it had certainly limited the extent to which they might be bankrolled.[10] Some degree of commercialization of biotechnology would be necessary, it was soon acknowledged, "in order to secure the system" itself. "We have the technology," Julio Delgado told *Time* magazine in 1996, "now we're looking for partners with money."[11]

For some, the way they went about doing this soon became detrimental to the nature of Cuban research. The priorities were always "short term," former vice director of the CIGB José de la Fuente complained.

> After 1990 fewer hours were worked, when you have fewer resources, you cannot expect scientists to have the same motivation as before.... The nineties were different in terms of the relationships between people. Inside the CIGB you now have a situation you didn't have before. Resource allocation became based on political aims and this affected the way people interacted.

Outside the CIGB it was even worse. . . . Each of the institutions wanted a way of securing greater access to funds.

He elaborated on some of these problems in an interview.

SRH: What was the situation like by the early to mid nineties, from your perspective?

JF: The country was in need of capital to maintain the project, and the rest of it. It was understandable in the way the whole project needed a way of obtaining the resources. . . . what was not clear or understandable for many of us . . . [was that] the only thing that mattered were projects related to exports. The result was a slowdown in basic research.

SRH: Were there any criticisms of this actually voiced?

JF: Strong criticisms were made by the scientists. It was discussed at all levels but the officials were not interested.

SRH: How did the political and economic difficulties of the Special Period change what was happening at the CIGB?

JF: It had a dramatic effect. There were few resources in the Special Period. This meant that at the center [CIGB] we were mainly working with what we had in stock. There was a need to make these projects pay. I certainly felt the need to develop projects that would resolve in a short time; I felt the pressure. The result was that all the more basic science-oriented projects of that time were given a low profile. Some of us discussed keeping them but that was not what the government wanted. So all the resources were allocated to these more applied type projects . . . and big projects like the anti-AIDS vaccine. Personally I was always opposed to the AIDS project, as I knew it would lead us nowhere—but he [Castro] needed something to show his presentation [sic]. . . . Of course: how else would you gain support for something many would think of as crazy at such a time? He had to stake a claim on the future possibilities and as often this meant taking the idea that something is technically feasible and then claiming that as meaning that it is more or less definitely going to happen, it is just a question of time.

SRH: What were the consequences of this for the CIGB?

JF: The CIGB now deals with research, production, and commercialization. That introduces a completely different way of organizing activities in the center. . . . For example, we all agreed we needed more R&D capacity. And we all discussed how to keep doing research despite the difficulties—there was no doubt about the continuation of the *development* side—the scientists [project directors] always discussed and understood the need to keep the basic research project going . . . as scientists, but that was increasingly not the main emphasis.

The main emphasis was to make science pay its own way, and that meant more easily commercializable projects were looked on with considerably greater enthusiasm than before. One such project was based on recombinant applications of erythropoietin, a hormone used in the creation of red blood cells and so crucial in kidney dialysis for chemotherapy patients. In Cuba, both the CIGB and CIM worked separately on their recombinant applications, with those countries not covered by the major Western pharmaceutical companies (particularly China) specifically in mind. But perhaps the best sign of the times, and certainly of a certain commercial drive, was to be found in the Cubans' development of Policosanol (PPG), an 8-alcohol extract of sugar cane wax. Developed in 1991, PPG is an original, Cuban-produced cholesterol-lowering drug that is well tolerated, even in elderly patients.[12] While it would evidently have some use in Cuba where the fat-rich national diet contributed to high levels of cholesterol, it was also very much geared toward the export market. And as many a Cuban scientist will point out to you with an ironic smile, PPG is an acronym in Spanish for "money making product" (producto para ganar). In both cases, however, potentially lucrative projects foundered not on the basis of the science but on the inexperience of the Cubans in marketing their products.

And it was not simply a problem of technical, economic, or political calculations—a problem of how to market drugs in a different regulatory environment, say (though, as we shall see, this posed its own problems). By connecting up with the global economy, Cuba's pharmaceutical regime was also brought into contact with the different cultural emplacement of biotechnology and biopharmaceuticals in the West. In rich countries, biotechnology has become a principal element of an individualistic conception of health. Under such a conception, the management of health is increasingly understood as part of one's "moral responsibilities to be fulfilled through improved access to knowledge, self-surveillance, prevention, risk assessment, the treatment of risk, and the consumption of appropriate self-help/biomedical goods and standards."[13] This is of course very different to what we have seen of the cultural foundations of biotechnology in Cuba. At stake in this attempt to reconnect was also therefore the particular approach to health that is a part of Cuban social and political practice more broadly. What had thus been *posed* to Cuban scientists as a problem of political-economy had therefore first to be *answered* as a far more generalized problem of epistemology. If Cuban drugs were to work in the world, the world would first have to be made to work in those drugs.

## Communizing Cancer

The Center for Molecular Immunology (CIM) is the newest addition to the Science Pole. It looks more modern, more compact than the other institutions. Like the CIGB and the Finlay Institute, it is a research and development institution. And, also like them, it works not only on drug discovery and research but on the production and scale-up of scientific developments too. Though not officially founded until 1994, the CIM has been in operation since around 1990. Its purpose is to continue the research into cancer therapeutics that had previously taken place at Cuba's National Institute for Oncology and Radiology (INOR), the general counterpart of the National Cancer Institute in the United States. The actual research projects and most of the scientists were in large part simply transferred from one to the other. Among them was the CIM's director, Agustín Lage, who had previously been deputy director of research at INOR. Before receiving his PhD from the University of Havana in 1979, Lage also worked in France, at the Pasteur Institute under Luc Montagnier, who was the first to isolate the Human Immunodeficiency Virus (HIV). Lage's commitment to foster innovative research displayed all the characteristics of the "ethically colored maxim" discussed in the previous chapter, and he was to play a central role in the CIM's success in the late 1990s. His passion, and accordingly that of the CIM, is cancer research.

*Cancer* is the umbrella term given to what are in fact hundreds of different diseases with one property in common: uncontrolled cell growth. The history of cancer research reveals almost as many approaches.[14] Before the Second World War, research was focused on tumor etiology and transmission—that is, on how cancer cells mutate and on whether they could be transferred from one organism to another. The big question behind this was whether susceptibility to tumor development was genetically inherited or not. Then immunological questions became the driving force of research in the 1940s and 1950s—research that used inbred animals to examine whether an immunogenic response could be provoked by chemically inducing tumors in animals. In some part these different approaches were framed by the nature of the experimental systems to hand. The arrival of standardized experimental systems such as genetically engineered animals made it possible to reliably test for genetic variations as linked to cancer. On the basis of these animal technologies, the sorts of questions that were asked from the 1960s solidified around genetic questions because that was ultimately what was best controlled for in these experiments. So it is that we come to know the world through the experimental systems that we use to depict it.[15] But what controls for these experimental systems?

Cornelius Rhoads, director of the Sloan-Kettering Research Institute in Manhattan—one of the United States' foremost cancer research institutes—once wrote "If one is to have a productive career in science, one must have some well-defined objective, whether this be the development of a better engine, the splitting of the atom, or the discovery of a better means for the control of cancer."[16] In short, scientists ought to work on goals that were "realizable," which meant they ought first to consider problems that were "doable." Joan Fujimura develops this notion of doability in her book *Crafting Science*. Itself a sociohistory of genetic research in cancer, she understands the construction of doability as central to the history of the specifically American cancer research that she examines. Doability means, on one level, answerability—as one molecular biologist is cited as saying: "'So with recombinant DNA technology you can ask the question "Are there changes in cellular proto-oncogenes in tumor cells?" and you can answer that question' rather than the far more nebulous question, say, of how to cure cancer."[17] But, on another level, it involves the articulation, or bringing together, of technological resources, experimental systems, subdisciplinary norms and procedures, even research questions themselves, all of which make something doable.

It is this latter aspect of doability that has most bearing on the Cuban case, for these doable problems contribute to what Fujimura elsewhere calls "accepted knowledge" or the "Central Dogma" of cancer research in the United States. In this case, the emergence of research on proto-oncogenes became, as she describes it, paradigmatic, a bandwagon to which everyone had to somehow hitch their work in order to make *that too* doable.[18] The construction of doability is thus in part about the determination of what does and does not get done in research. Briefly put, in the West, scientists came to know cancer not only within the confines of particular experimental systems but within the conceptual architecture of the doability paradigm also. Those research agendas that most nearly conformed to this guiding rationality—such as proto-oncogene research, whose particular strategies and methodologies focus attention on the genetic basis of cancer and so *explicitly* seek to exclude the background noise of other data on, say, the viral properties of cancer—were thus the more doable and thus were the ones that got done. Fujimura considers that such articulations of scientific knowledge and power might be either constraining or emancipatory, but in this chapter I want to inquire into the unevenness of how these constraints and possibilities are apportioned, and the consequences of their so being.

## The CIM and Cancer Research in Cuba

Cancer accounts for most deaths among people under sixty-five in Cuba. For those over sixty-five, it is the second cause of death after heart disease.[19] As discussed in chapter 2, to some extent this meant Cuba had a first-world disease profile, characterized by the predominance of chronic over infectious diseases. There is a difference, however; cancer is not everywhere the same. According to the WHO, as much as 23 percent of malignancies in developing countries are caused by infectious agents (including hepatitis B and C (which cause liver cancer), human papillomaviruses (causing cervical and anogenital ulcers), and helicobacter pylori (causing stomach ulcers), as opposed to only 8 percent in the developed countries.[20] If the Science Pole was to contribute to national need, therefore, then here was a pressing area that previously had been largely overlooked in the work of the CIGB, Finlay, and the other larger institutes there. It was work that would be taken up most fully at the CIM.

It was the research group under Lage that formed the core of the CIM's work in the mid 1990s. It consisted of around thirty scientists, some with medical backgrounds, and others who were physicists, chemists, biochemists, and engineers who had been working together for over ten years; there was only one biologist. The work of this group affords a unique vista on the actual conduct of scientific research in Cuba; it gives an indication as to how the forms of social organization described earlier corresponded to a particular sort of epistemological practice. Given his position and his passion for writing about Cuban science, Lage's account of this group's work further helps us to understand how the Cubans themselves understand what it is they are doing in their clinical and laboratory practice.

That practice was primarily focused on a particular subset of cancer research: immunotherapy. Physicians have been aware that the immune system might play a role in controlling cancer for more than two centuries. In 1774, a Parisian physician injected pus into the leg of a patient with advanced breast cancer and observed that the cancer improved as the infection worsened. In the early 1900s, American surgeon William Coley treated over eight hundred patients with toxins to reduce their cancers. While the data remain controversial and have not been reproducible, for a long time they nonetheless indicated an important potential means of reducing tumor growth.[21] Today, the basic idea of cancer immunotherapy remains to try and somehow harness the body's own immune system to reject cancerous cells. But the basic problem with this idea is that cancers usually arise from host cells themselves, presenting more or less the inverse of the problem of organ rejection, where transplanted organs are recognized as foreign and attacked by white blood cells. Attempts to mobilize the immune system to reject cancers thus have to overcome the immune system's innate tolerance of its own cells.

It is heavily ironic that the Cubans would thus take up a branch of cancer research that had to date been most consistently marginal to the routine approaches to cancer: surgery, radiation, and chemotherapy. As Stephen Hall's history of cancer immunotherapy *A Commotion in the Blood* reveals, immunotherapeutic approaches have usually been seen as the least likely, the least doable of approaches to cancer. As a result, they have not garnered many of the billions of investment and tax dollars made available to cancer research over the past century, nor have they warranted the interest of the field's top specialists. So whereas the Cubans chose to develop interferon off the back of that drug's hype in the West, in taking up immunotherapy in cancer they were positioning themselves well away from the cutting edge, focusing on an area that was still largely untested. If such a choice was ironic it was not, however, unpredictable.

Research on cancer at Cuba's National Oncology and Radiobiology Institute (INOR) was initially centered on a group of theoretical biophysicists modeling antigens and cellular receptors related to some of the diseases and cancers affecting the Cuban population, and one of the early successes of this work laid an important foundation for Cuba's approach to cancer. In 1982 the Cubans developed a diagnostic tool known as the Ultramicroanalytic System (SUMA) which provided an effective, domestically produced alternative to the more expensive ELISA technology still commonly used in diagnostic kits in the West. At the heart of this technology were monoclonal antibodies. Monoclonal antibodies are clones of a single antibody cell, first developed by American researchers Georges Kohler and César Milstein in 1975 (for which they were awarded a Nobel Prize in 1984).[22] The discovery sparked considerable excitement in the late 1970s and into the early 1980s that these too, like interferon, might be the magic bullets to beat cancer. The first Cuban monoclonal antibodies were produced in 1982, and they were given the go-ahead for clinical trials in 1984. By 1989 they were being clinically applied in Cuba via injection as a treatment for organ transplant rejection. The Cubans' initial interest in them was thus rather more practical. Monoclonal antibodies facilitate a range of medical practices—including complicated transplant operations—

that would otherwise simply be impossible on the island. And once they knew how to produce them, the Cubans began looking for other possible applications.

As a result there was always in Cuba plenty of what, elsewhere, was a rather rare resource, offering generous scope to experiment a little with this technology. The CIM now produces "several kilograms" of purified recombinant proteins and monoclonal antibodies per year, in hollow fiber and stirred tank fermenters operated from the control room next to where I had met with Pérez previously. Our discussion had taken in the history of the use of monoclonal antibodies in Cuba: he tells me they were the principal technological platform at INOR and its successor institute, the CIM. So where American cancer research had its industrial research model promoted by postwar industrialists like Jewett and Whitney, and taken up in institutes such as Sloan-Kettering in New York, the Cubans had opted to specialize in an area they gambled on being potentially highly rewarding, but which also afforded the recompense of being practically useful in other domains in the first instance.[23] And as with interferon before, this work progressed rapidly from research through to clinical application.

The initial use of monoclonal antibodies in Cuba then was as an enabling technology to support the government's promise of advanced health care. But monoclonal antibodies also offer the potential for delivering targeted therapy: the idea being that monoclonal antibodies can be engineered that will bind to specific cancer cells and potentially render them inactive. As was beginning to be examined elsewhere, in the early 1990s, researchers at the CIM likewise came to develop monoclonal antibodies for just this purpose. The Cubans turned to this work with a different approach to that being pursued elsewhere, however.

The experimental system with which the Cubans sought to use monoclonal antibodies as a means of delivering targeted cancer therapy was based on two key differences: one material and one conceptual. Materially, the production of large quantities of monoclonal antibodies in Cuba and their use in "mundane" diagnostic technology made an approach based on therapy rather than cure rather more "doable." Conceptually, their understanding of therapy and treatment was itself derived from their prior greater experience in infectious diseases research, something that tilted them towards a more system-wide and interdisciplinary approach.

On the basis of this theoretical architecture, something else took place along the trajectory from the more industrial work of antibody production to the elaboration of complex therapeutic models. This becomes clear by looking at how this work at the CIM developed from the Cubans' earlier attempt to understand a particular component of the "cancer system"—epidermal growth factor, or EGF.

## The Problem of EGF

In 1984, Lage and his colleagues at INOR were the first in the world to describe the role of EGF receptors in breast cancer, in a paper in the journal *Breast Cancer Research and Treatment*.[24] EGF is a cellular protein that plays a key role in stimulating cell growth, binding to cells via epidermal growth factor receptors on the cell surface.[25] Lage's group had found that epidermal growth factor receptors (EGFRs) were overexpressed in 60 percent of human breast tumors.[26] At this time, only a few other groups had done much work on EGF in relation to cancer, but the group under Lage would soon specialize in it. In a 1986 article in the journal of the Cancer Research Institute of the Slovak Academy of Sciences, Lage, Pérez, and Jorge Lombardero reported on further studies carried out on the epidermal growth factor on mice. "Epidermal growth factor was rapidly distributed, reached tumor cells and recognized specific cell membrane receptors," they announced. The implications were significant. "These results suggest that high doses of EGF could eventually be used for inhibition of the cell proliferation in some tumors."[27]

"Until now," wrote a subsequent editorial in the American-based magazine *Bio/Technology*, "no one had seriously considered that saturating these over-expressed [EGF] receptors with EGF alone might have clinical utility."[28] That is not quite true. Other groups had considered EGF (and the work on EGF receptors would be taken up in many other places). But it *was* the case that EGF had on the whole been seen as part of the problem—to be tackled by targeting the EGF receptor so that free-floating EGF in the body would be prevented from docking with and nourishing tumor cells—what in the literature is termed a passive approach.[29] What Pérez and Fonseca were proposing, by contrast, was to use human EGF as part of the solution, that is, as an *active* agent with which to interfere with the normal (but cancer producing) binding of EGF to its receptor. This approach potentially offered the advantage of requiring just one or two doses, and of being cheaper to develop. But it was higher risk and would, at the end of the day, offer less capital gain because it would not be administered as a lengthy course of therapy. In many respects, therefore, this made such a specific-active approach rather the opposite of what the global pharmaceutical industry might have wanted.

The research landscape the Cubans were embarking upon—that of cancer immunotherapy—was far from percentage science, therefore. As one group of scientists have put it: "The results of decades of work, thousands of patients, and millions of dollars can be summed up as follows: Almost nothing worked, but very occasionally a good response was seen for reasons that were not clear, providing enough incentive for the whole field to keep moving." The Cubans believed that they might just have found a way of inducing a good response, however, and regardless of the high chance of failure, they *did* have certain advantages to bring to this work—not least their wider experience with some of the adjuvants and diagnostic technologies that would be necessary to making it work in practice not just in theory. An active immunotherapy approach was thus rather more conceivable in Cuba than elsewhere, and this was at least in part because of the way that Cuban science was positioned outside of the dominant do-ability paradigm of cancer research in the West.[30]

For anything to come of such an approach, however, the Cubans would also have to overcome a number of equally challenging scientific problems. Even within the still developing field of cancer immunotherapy, they were embarking on a little trod path. Injecting human EGF to induce an immunological response that might interfere with the way that the body's own EGF nourishes tumors, was an approach with little experimental history behind it. The Cubans' work was thus at the very margins of the already marginal. And before any progress could be made, they would need to confront the central problem in this line of research: tolerance to cells that the body believes to be its own, or what is known as "self-tolerance." If EGF, which occurs naturally in the body, was to be used as part of an active immune approach to cancer, how to overcome the remaining problem of tolerance to self-EGF?

### *The Normal and the Pathological*

The immune system's difficulty in recognizing cancer as foreign to the body is part of what makes cancer so stubborn. Self/nonself recognition is thus the central problem of cancer immunology, and for the researchers at the CIM who wanted to develop an active immune approach using EGF—finding a way of "training" the body to respond to EGF and so produce cancer-specific antibodies—was a hurdle that had to be overcome. So far, no cancer researchers had managed it, and the reason may well be that the switch from a passive to an active (targeted) approach required a different sort of expertise, one that for the Cubans at least was available not so very far away, in the therapeutic experience of the vaccinologists at the Finlay Institute. In fact, the researchers at the CIM would soon come to share not only in the work done on meningitis B at the Finlay Institute but also the subsequent development of that work at the CIGB, where a team had produced a recombinant protein from the *neisseria meningitides* bacteria, P64 K. What they would take from this other work would be telling. By using P64 K as a carrier protein with which to introduce EGF into the body of patients, the researchers at the CIM were able to find a way to break tolerance to self-EGF.[31] EGF now disappeared as the root of the problem, which switched to being one of finding even more efficient adjuvants to stand in place of the P64 K protein. The delivery problem had displaced the problem of EGF. A small scientific advance had occurred. An active immune approach to cancer had become a little more doable.

The logic behind all this unfolded later, when Agustín Lage commented:

> It is not to be expected that, acting on just one molecular component, we could shift the whole [immune] system out of its tolerance attractor. Natural tolerance is a product of organisms, not of cells or molecules, and as we move beyond particular cases in which one antigen plus one adjuvant do the job well, we will need "system biology" approaches to the problem of active specific immunizations.[32]

The point is that such a holistic view of the body—which today is rather more in vogue—was most readily mobilized or, rather, experimented with, in Cuba in the early 1990s on the back of an equally holistic material or practical enframing of scientific work itself. By expanding the ways of making EGF doable, they found other—more fundamental—ways of "doing" EGF within existing therapeutic models. As John Gribben of the Cancer Research UK's medical oncology unit noted, "We might even ask how much the doability factor now impedes our ability to move things forward [in the West]. It isn't always a question of saying they [countries such as Cuba] have set the bar too low; it may be that we have set ours too high."[33]

To some extent it is tempting to understand what the Cubans have been doing as a sort of socialist epistemology of the body, and one might find certain grounds for thinking so. Immunotherapy in Cuba wasn't really seen as a technoscientific fix to cancer because the Cubans were not looking for such a fix. Through their work on EGF—the conceptual underpinnings of which would later be taken up in a more wide-ranging set of immunotherapeutic approaches, involving monoclonal antibodies—they were instead looking for a way of helping the body to help itself, as it were. They were looking for a way of enhancing the relations between the various biophysical mechanisms of the body. "The moment which must be considered strategically decisive . . . in a history of the biological sciences," wrote Michel Foucault, "is that of the constitution of the object and the formation of the concept." And here were both in action. The constitution of EGF as an object of cancer therapy was not unique to the Cubans. But the formation of a new theory about how to approach it was. It involved a conceptual shift from EGF as the *problem* (something which nourishes tumors and yet is natural to the body and so must be blocked from being taken up by tumor cells) to EGF as the facilitator of a new mechanism (an active immune response in which the body might be "trained" to recognize self EGF). It was a question of turning a problem into a tool. And it rested on a willingness to conceive the pathological as but a part of the normal functioning of the immune system. All this was rather different to the dominant trend in Western cancer research, where EGF could be said to have been seen as part of the problem of cancer and not a part of the cure.

The way that EGF was positioned within Cuba's *political economy* of cancer research therefore shaped the way it was understood as a problem within the *science* of cancer research; this in turn rested in part on a socialized definition of the self-other problem. In short, in the West, EGF presented a social barrier to cancer research *before* it presented a scientific barrier. Under the different social conditions for science in Cuba, the problem of EGF was differently configured: in fact it came to be considered the very opposite of a scientific problem. It came to be considered a scientific possibility. The intention here is not to raise the Cubans' work on cancer immunotherapy to the level of some sort of originary moment indicative of a broader paradigm shift; the point is to understand how it was that such a shift in thought about cancer came to be articulated *in Cuba* as opposed to anywhere else. But if reconfiguring the problem of EGF had, we might say, communized cancer somewhat, this too was only part of the problem solved.

## Gaining Warrant

By the mid 1990s there were more than a few foreign companies willing to risk investing in some of Cuba's often quite innovative science. Such investments are very much part of the normal operation of biotechnology science in the West. That they began to look at the work going on in Cuba was in part because of the Cuban government's promotion of its biotechnology efforts and in part because of the Cuban scientists' oftentimes fresh approach to intractable or sticky scientific problems. One of the first companies to become properly involved was a Canadian venture capital firm, York Medical (now YM Biosciences).[34] YM Biosciences is a company founded on the problem of cancer. As CEO David Allan himself describes it: "We focus on the disease and then try to find the treatments that might work best." YM Biosciences is, properly speaking therefore, a cancer company. Its interests are, properly speaking therefore, closely in line with those of the CIM. And cancer therapeutics suit a small-scale merger of this sort. The aim for a company such as YM Biosciences—which does none of the science itself (though it will have scientists on its board of directors to offer peer review oversight of investments)—is to in-license promising compounds from what are often quite small biotech outfits. It then takes those compounds through the various stages of development needed to turn them into marketable drugs: clinical trials, obtaining regulatory approval for their sale in the markets to which the company has rights, partnering with relevant companies, and so on.[35] If successful, the company will then try to out-license the drug to a larger pharmaceutical company for production and marketing.

Companies like YM Biosciences fill a growing niche that has been opened up as patents expire and the "big pharmas" who formerly owned them look to cut costs. Some believe, therefore, that companies like York Medical are in the driving seat of changes afoot in medical practice.[36] But for that they rely most crucially on the perceived benefit of what are always experimental drugs and, as we shall see, such credibility in contemporary bioscience is derived not simply from the validity ascribed to scientific claims. So too must those scientific claims obtain economic and political efficacy, and this turns as much on making local scientific systems credible as it does on making the findings produced by them believable. In short, it turns on precisely the fault line between the bioethical domain in Cuba that supports the sort of reformulation of research paradigms seen above and the various paradigmatic forms of regulation of expertise germane to the global pharmaceutical economy. For the Cuban cancer drugs to gain a market share and thereby make it into bodies outside of Cuba, they would first have to prove themselves against these warrant-making mechanisms.

### *The Clinic on Trial*

I got into a discussion on this very topic one evening with a medic from the Hermanos Ameijeiras Hospital, which stands squarely against the sea front overlooking the Malecón. The Hermanos Ameijeiras Hospital is the largest of its kind in Havana. It carries out many of the clinical trials for the institutes of the Science Pole. As we stroll in front of the partially renovated buildings which line the Malecón, she introduces me to some background on these trials. In Cuba, she sees two dominant ways in which notions of bioethics have been deployed historically and which inflect on (or help make possible) Cuban scientists' particular approach to their scientific work: the practical and the political. Ernesto Bravo, biochemist and director of NACSEX, is also the author of a series of published interviews with key figures from the world of Cuban biotechnology that sheds some light on the first of these, the practical. Bravo notes that a major factor that assisted the development of Cuban biotechnology was that "there were no institutional conflicts or polemics regarding bioethics as happened in other countries"; there was no debate. By this he means the state set the standards by which drugs were measured.

During another discussion with a health specialist, some light is further shed on the political. "Ethical considerations are very important in Cuba," my interviewee says. "The clinical trials done here are taken very seriously, they are very responsible and they don't fall into the problems that we are seeing in the world today which prioritizes the marketing of products and not really the value that the product may have for human health." It is a point of view baldly stated, but as he acknowledged, the problem now is that Cuba is no longer isolated from the West. I am told this during an interview that took place against the background of a diplomatic wrangle that was then unfolding between Uruguay and Cuba. The interviewee claimed that Cuba was still giving 2 million doses of meningitis B vaccine to Uruguay free of charge, despite the Castro government having been criticized heavily by the Uruguayan government. Another Cuban social scientist mentions the same a few weeks later when I put these claims to her that Cuba gave the vaccines regardless of the political split. "This for me is bioethics," she said. "When there was this break of relations with Uruguay, Cuba continued providing, free of charge, its vaccines for the children, regardless of the diplomatic conflict." Such understandings of bioethics in terms of the broader redistributive mission of the Cuban revolution are interesting for the way that they are themselves underpinned by the perceived moral legitimacy of Cuban biomedicine.

These observations all help to explain why clinical trials had not presented much of an issue for Cuban biotechnology in the 1980s. Not least, the first clinical trials (those associated with the meningitis B vaccine, for example) had been organized and carried out by the same centers that produced them in conjunction with the Ministry of Health. As the same vice director at Cuba's National Center for Medicine Control (CECMED) who told me this recalled when discussing the meningitis B project, "The first trials on this product [VA-MENGOC-BC] and on others had finished before the creation of CECMED even. Those trials were controlled by the Vice Ministry of Research and Education at the Ministry of Health." More recently, when there have been no appropriate national standards (often the case in the early 1990s), the various centers—in liaison with the emergent CECMED—would establish an agreement among themselves that set the relevant WHO directives as the benchmark for quality control assessment. This differs markedly from the formerly dispersed and ad hoc nature of clinical trials, she noted. Indeed, arguably, it closed a window behind the VA-MENGOC-BC. "Ten years ago," another colleague at the National Center for Clinical Trials (CENCEC) said, "when we went to the country to conduct clinical trials, people didn't know what they were. Now most people do." And the shifting moral terrain encompassed by the island's reintegration with the global economy was to alter that milieu even more noticeably. Fast science had just run into rather sticky ground.

The modern randomized clinical trial has been a major influence in biomedicine since its widespread adoption in the 1960s.[37] Clinical trials involve a range of work, from securing regulatory approval, to selecting sites for the trial to take place, to developing protocols monitoring the trial and analyzing the data. What is at stake in all this is the "value" to be assigned to new biomedical knowledge. Through warrant-making forms like the clinical trial, what biomedical knowledge "counts" is no longer determined solely by the veracity of the science (the effectiveness of a particular approach to EGF, say); now it is determined by what that knowledge "does" (and this is strongly linked to the doability issue we encountered earlier) within particular domains (such as a given disease market, for example). In other words, what sorts of cancers was EGF likely to help, and were these cancer treatments the most profitable? Under a neoliberal pharmaceutical regime, one has to make a market at the same time as one makes a drug. Whereas previously the Cubans simply wanted to prove the efficacy of their drugs, now they had also to prove the desirability of such drugs, and they had to do so without compromising that productive space of operations that was predicated precisely on a certain *distance* from Western science. They had taken the science as far as they could go. The as yet unsanctioned EGF vaccine stood before history. It existed but had yet to be brought into the world.

This is where a company like YM Biosciences comes into play. It is very much their role to strategize all this. And as CEO Allan sees it, in this not only does YM Biosciences fit a vertical niche, as many business analysts have pointed out; it fits a horizontal, or geographical, niche too. YM Biosciences is able to coordinate the development of products across diverse territories and regulatory jurisdictions. It was precisely the sort of company the Cubans needed. When we speak, Allan is not shy, therefore, in publicly discussing the company's real competitive edge: *locating* products is "the corporate secret." In part that turns on being "highly visible" within the corporate "community," attending all the biopartnering and investor events that make up the corporate biotechnology "circuit."[38] In part, it also turns on access.

### *Experimental Markets*

Emerging markets can be a lucrative resource for investors, but they hold the risk of the unknown and, most of all, the unpredictable. York Medical first struck on what the Cubans were doing when the company was still in its first incarnation as Yorkton Securities, an outfit that had been investing in mining in Cuba for some years. One of its analysts was told of possibilities in the biotechnology sector, and an exploratory visit was soon paid to the island by a small team from the company, headed by Allan in his role as nonexecutive chairman. Allan professes to having been amazed at what he saw, and he considered the commercial opportunities that the Cubans' work represented worth the not inconsiderable risks. Behind the scenes, the directors at York Medical spent time building up trust with the Cubans. Always keen to further Cuba-Canada links, former prime minister Pierre Trudeau himself got involved. Castro was wined and dined, and the Canadians proposed an extensive collaboration.

It was a timely offer for the Cubans. By the mid 1990s, the biotechnology centers were, as we have seen, in dire need of capital investments. Initial sales of the VA-MENGOC-BC had largely been realized by this time and little financing was available from elsewhere. Because of the embargo, American companies were prohibited from investing in the island's biotechnology endeavors and, for the same reason, European firms remained hesitant. Against this background, the State Council was eventually convinced that the Canadian deal offered the best way forward, and the go-ahead was given for the first foreign investment in Cuban biotechnology. York Medical immediately set about allying Canadian experts in pharmaceutical licensing, regulatory and clinical affairs, and marketing to the top players in the Cuban life-sciences establishment: Limonta (still then head of CIGB), Lage, and Carlos Gutiérrez (then head of CENIC).[39] In the process, a new relationship was put into place that would play a significant role as Cuban biotechnology confronted the forces of global biocapitalism at the very moment when the regulatory structures of the pharmaceutical industry were tightening up considerably. It was a precarious moment for the Cubans to be venturing on to the world stage, but for the same reasons it might just be propitious.

The enthusiasm at York Medical was matched from the Cuban side. In 1995, an Office of International Collaboration (DCI) was established in Cuba in order to "promote and facilitate business and cooperative investment in the fields of science, technology, and the environment and the peaceful use of nuclear energy."[40] It marked an important turning point in the willingness of the Cuban state to negotiate with foreign capital. At the same time, legislation was proposed to allow joint ventures to be formed more easily. The Foreign Investment Law (Law No. 77) passed in September 1995, defining clear procedures for investment applications and providing certain guarantees against expropriation of property—a persistent fear among investors in socialist countries. This law also allowed for the possibility of investments of up to 100 percent.[41] The necessary window for York Medical had just opened and, as the York Medical collaboration got into full swing, Cuba's "biopharmaceutical industry" (the name change being significant) was confirmed in a business review of February 1996 as "one of the country's strategic pillars for economic and social development." Two years after the first meeting with the Cubans, the York Medical team eventually "cherry-picked"—as the company literature had it—five key products from different market segments "to represent Cuban potential to the largest global pharmaceutical and medical concerns."[42]

This was a promising start. Armed with its Cuban products, York Medical was now not only the major representative of Cuban biotechnology, it also had "one of the broadest cancer portfolios in North America," as a representative in Havana told me. Its strategy was now to in-license Cuban biotechnology products, adding value through its management expertise in return for a share of the products if and when they reached the market. In the meantime, the company planned to raise capital (a good chunk of which would be invested in Cuba) through occasional equity issues. But as the blurb on "forward-looking statements" attached to any investment literature states very clearly, many things can go wrong in a business of this sort. And not least among the obstacles to immediately confront York Medical was that raising funds took longer than anticipated. In a management shake-up, a number of executives left the company and, at year's end, a $20 million offering fell through in Canada, largely because of the company's Cuban ties. York Medical was forced to settle for $12.5 million through a private placement.[43]

## *Epistemic Resistance*

Changes were afoot on the Cuban side too. By the second half of the decade the worst of the Special Period seemed to be over, and suddenly commercialization was not quite the priority it had been. In 1997, the Fifth Party Congress concluded by endorsing economic reforms (including financial and enterprise reform) and in particular tightening financial management in those industries, such as biotechnology, that used hard currency or imported inputs. Admittedly, on the surface the concern to open parts of the country's economy up to foreign trade was confirmed as an essential priority, at least in the short to medium term. And interest was expressed in building on the $2 billion raised via joint ventures so far.[44] But at the same time as this critical capital was being snorkeled in, "substantial" changes were also made to the country's industrial structure. The overall economic packaging of biotechnology, for example, was now to be overseen by Cubanacán, the state-owned company better known for managing tourism. With seventeen companies in its portfolio and $600 million in assets, Cubanacán had interests elsewhere. It saw tourism, the sugar agro-industry, and nickel mining as the way forward.

What just a few years before had been a prioritization *of* biotechnology slipped to a prioritization of "particular strategic areas" *within* biotechnology, agricultural applications among them.[45] Efforts to make biotechnology profitable for the country now began to focus on the implementation of a program of import substitution and domestic production of drugs which emphasized the need to "better satisfy" the basic needs of the population. The merger with York Medical would continue to go ahead but it seemed an about-face had taken place: biotechnology was no longer a priority for commercialization. Reflecting this, the directors of the various institutes of the Science Pole were reminded of biotechnology's public health roots. For many of the institutes, generic production was being put back on the map. Little comment was passed on this either at the time or in our subsequent interviews. But the problem, it was noted, was that the new commercial priorities had exposed the lack of modern management techniques within the centers.[46] The Cubans were aware they were getting hopelessly out of their depth.

The outcome of the ACC's strategizing was announced in a *Dictamen* on the Strategy for Scientific and Technical Innovation (the date coinciding with the thirty-sixth anniversary of the founding of the National Commission for the ACC and the beginnings of that body's rejuvenation under the revolution). The news was cold. Rosa Elena Simeón, former head of the ACC and then minister for Science, Technology, and the Environment, stated that biotechnology was not open to direct investment. Only joint venture alliances would be considered, she announced in a less-than-clear statement: "We cannot open up to direct investment a field whose development has cost us so much in human and material resources, but in the sale abroad of the results of these efforts, it is often essential to rely on the help of entities or persons with know how and prestige in a very sensitive and specialised market."[47] Another problem is that revenues for biotechnology over the previous three years had fluctuated from $102.3 million in 1996 and $290 million in 1997 to $50–70 million in 1998. Revealingly, when the ACC hosted a team from the American Association for the Advancement of Science (AAAS), the team toured the biotechnology facilities and established contacts for future scientific exchange, but there was little talk of actually doing business.[48]

York Medical was by now committed though and had to plough ahead. In February 1998 it invested Can$7.5 million on Clinical Trials in Cuba, Canada, and Europe for its key Cuban products, particularly DiaCIM, a radiolabeled in vivo cancer diagnostic test.[49] Including York Medical's portfolio, around thirty-five new biotechnology products from Cuba were expected to hit the market the following year. Accordingly, that year's annual conference in the Science Pole, "Biotechnology Havana," was subtitled "From the Laboratory to the Market." Exhibit space rented for $50 per square foot. Biotechnology was increasingly spoken about now as a "bet," a risky venture. On the basis of recent statements it was not at all clear whether that meant it was seen as an opportunity or a pitfall. Perhaps one answer lay in the fact that "professionalization" remained a major preoccupation within the Science Pole. The vice minister of public health, Abelardo Márquez, called for a strengthening of capitalization and again for the adoption of modern management techniques within the biomedical centers, for example.[50] Professionalization, it seemed, had by now become a sort of discursive umbrella under which capitalist practices such as the taking on of risk capital could be incorporated within what otherwise appeared to be the continuing socialist-oriented practice of Cuban biotechnology.[51] The problem, as reflected in the policy reversals and changes of heart on both sides, was often represented as a clash of cultures: an inability to fully understand another system. It is often understood as this even today—a recent article by Lage attests to the particularities of the Cuban approach being "hard to explain." But the problem is more pervasive than this: what was ultimately at stake—the Cuban drugs York Medical was seeking to develop—were themselves the product of *both* these different rationalities. It seemed that it was proving harder to get EGF to bind to capitalism than to cells in the body. But, again, a holistic view would provide the solution.

## *Locating the Problem*

By 1999, York Medical's business plan was two years behind schedule: held up by the delays in transferring rights, gaining credibility for Cuban products, and repeating some of the clinical trials. Held up, that is, by precisely the additional work required to bridge these two systems of thought. In January, David Allan became the company's new CEO. He believed that more gains were to be had by focusing on fewer products with greater potential, centered in particular on cancer therapeutics. His views sat well with the changed situation in Cuba and meant he was prepared to continue the company's close involvement with the work being done at the CIM. As Allan saw it, the CIM was conducting both original and promising research that needed to be far better exploited. With a newly delimited platform, the two sides drew closer together once more. But there were still no guarantees. York Medical already had a history of losses, and clinical testing might fail to produce positive results. Even if testing succeeded, there were no assurances that any patents would be granted or that they would be sufficiently broad to crowd out competition. One of the collaborative partners might withdraw. The technology itself might fail.

Risk was the watchword then, but Allan's new strategy was timely. York Medical took these risks on with what Allan saw as certain strategic assets. First, York Medical was a relatively unique company simply because it dealt in products. Even two years later, at BIO 2001 (the major biopartnering event) less than ten of the more than eight hundred companies represented were actually offering products as opposed to services and platforms. As Lage comments in an interview: "Biotech is about manufacturing things. Most biotech in the world still can't do this." Cuba also gave York Medical two further advantages: first refusal on all its products, and, ironically, as a result of the 1996 Helms-Burton Act that further tightened the US embargo, a certain financial protection. No American company could buy York Medical owing to its partnership with the CIM. In turn, York Medical would make a significant impact on the biotechnology milieu in Cuba, and particularly on the development of the CIM through financing capital improvements and raising its profile internationally. Indicative of this, Lage was himself elected to the board of directors at York Medical, making him the first Cuban to be appointed to such a position. York Medical then invested $10,080,000 in CIMAB—the CIM's business wing—to be paid over the next five years.[52] In doing so it made not only a financial investment but an investment in the idea of a specifically Cuban biotechnology too, in which it now had a substantial stake.

Exemplary of this was the idea of representing Cuba to potential investors as a "company": Cuba SA. It was a simple twist of phrase, on one level. But it gave both sides a middle ground they could "buy into" from what would always remain quite different perspectives. It seems it was Allan's idea:

> We were struggling hard to explain what we were trying to do to them—they consistently had difficulty understanding why *I* didn't just make decisions for the company, and it was a different context or way to sell or plant the idea. And that's where it came from. We hit on the notion of saying our responsibility is to our shareholders and yours is to your shareholders, the people. You know I found they were making unilateral decisions. When I first started working it was with the government. They took great pride in the decentralization they had in their science. Ironically I said to them that you should *centralize* as the people in charge [of the different institutes] don't know about business planning, there wasn't proper consideration. So I said, "Don't you understand that each and every citizen has given up something to make this *possible?*" . . . So you have to take them into account, and that took seed and rooted from there. It started up from the pride they took in *decentralizing* decision-making in biosciences.

Whatever else it was, the metaphorization of Cuban bioscience as a national investment in which the Cuban people all had a stake carried considerable political currency, and it provided a necessary way of mediating between the competing epistemologies of capital markets and socialist medicine.[53]

### *Locating the Frontier*

So the Cubans had solved one of the problems associated with EGF by developing a scientific model that saw it as a vehicle for the delivery of an immunotherapeutic approach to tumor cells. And they had begun to solve the problem of making that science work in the world by finding a company to more generally help take their research through the various stages of capitalist drug development, from establishing warrant in clinical trials to securing economic support in the market. But now they encountered a third problem, because no one else did science quite like they did, and they didn't fully appreciate why this might be a problem. Allan did though. As he suggested:

> *Time* magazine had a four-page spread on [Cuban biotechnology] about four years ago, and through the work that we have done, we have articles in *Science* magazine and *Scripp* magazine, and then there is the work of the *Economist* conferences, and the round table that you have in Cuba. Not withstanding all that work, the level of appreciation of what's happening is extremely low, and why that is I just simply don't know. And consequently it is a huge burden for us to try, from a financing point of view, to be carrying the Cuban story around because first of all it is not known, secondly when it is known, it's because of its alienation from most of the rest of the world.[54]

Allan himself took up the cause in various company interviews, conferences, and business news appearances. But he most consistently presented the science as cleft of its particular provenance in Cuba, hitching it instead to the growing interest in cancer immunotherapy, which by the late 1990s was increasingly being taken up within global biotechnology. It was a smart move. By 2005 globally more than fifty vaccines had been put in for clinical testing and over four hundred clinical trials had taken place. Through these trials, understandings of cancer came to be rooted more fundamentally in immunology. The research "frontier" was moving towards the Cubans. A thread posted by a spokesperson for the CIM on the discussion group Bionet captures well why: "Molecular immunology is a field of interphases. It is a mixture of immunology, molecular biology, cell culture, fermentation, pharmacology, clinical expertise and so on. Success depends not only on how in-depth one can go in any of these disciplines, but also on how to mix knowledge across disciplinary and political borders."

## Sticky History

The collapse of the Soviet bloc and what some called the "end of history" was in fact for many a reminder of the very real importance of history and ideology in affairs, such as science, that are both intimately local and uncontrollably global. The Cuban breakthrough with EGF played out against these wider developments. In doing so it offers a useful perspective on the nature of originality in Cuban research. In Nietzsche's formulation, originality meant "to see something that as yet has no name, that cannot yet be named, even though it stands before everyone's eyes." We might say that what is required, in order to be original, is to bring an object into history. For a long time the Cuban work on monoclonal antibodies took the guise of a seemingly unoriginal craftwork, until it transpired that this work was itself a possible solution to certain newly posed questions about the role of EGF in cancer growth.[55] But, once posed on a socialist island in the 1980s, how was the *potential* role of EGF in inducing an immunological response to cancer to be brought into a history that, in the 1990s, had taken a decidedly Western, capitalist turn?

In his history of yellow fever research, a story set partially in Cuba of the late nineteenth century, François Delaporte compares two nationalist accounts of the discovery of the cause of yellow fever: the Cuban account, which postulates Cuban physician Carlos Finlay as the discoverer, and the American account which postulates the head of the US Army Yellow Fever Commission in Cuba, Walter Reed. It is one of the classic historical debates over scientific originality given a particular edge by the broader political rivalry between the two sides. The details of both claims need not detain us here, for in a brilliant analysis Delaporte reveals both Cuban and American accounts to be false. Indeed, far from there being one discoverer, what took place in the last two decades of the nineteenth century, he says, was an epistemological shift—a change in the very conceptual architecture of research, based on a diffuse set of studies that took place in locales ranging from Asia to Scotland to Cuba. These developments enabled scientists, collectively, to conceive of the mosquito as a vector for the transmission of disease. In so doing, they ushered in a new era of epidemiology.[56] The Cuban work with EGF is less groundbreaking than that described by Delaporte. But it does reveal how the sort of epistemological event he describes is profoundly spatialized.

The Cuban breakthrough, predicated on a long history of work undertaken at the periphery—reading the papers available in the less prestigious journals, working with what they had and in any way they were able—was realized when applied to a problem of the center that was not even raised as a problem *in* the center: the doability paradigm. The problem of the doability paradigm created the problem of EGF at the same time as it was concealed by it. It was the Cubans' geographical dislocation from this paradigm that enabled them to conceive a way around it: by thinking of EGF as a vehicle and not as a barrier to the treatment of cancer. Put more simply, what was believed to be a problem (and so produced as one) in one place was not a problem at all in another. Conceiving of EGF as a vehicle was but a first step, however.

That breakthrough also had to be brought into history, and it was York Medical that was best positioned both to see the originality of such work and how it might be materialized. York Medical's strategic expertise in the global marketplace enabled it to correctly identify the value of, and then to secure warrant for, this and other aspects of the Cubans' work on cancer, something that the Cubans had palpably failed to do on their own account with some of their other products (PPG, for example). But for York Medical's work to also succeed required both the Cubans *and* the Canadians to then jointly solve a further problem: that of inserting the Cubans' science in the global pharmaceutical economy without jeopardizing their strategic advantages: their being located "outside" of the system. And what enabled them to do that, most fundamentally, was that York Medical was able to sustain within *its* particular business model the idea of the researchers at the CIM continuing their original work under conditions of *their* own particular ideologies of science. The history of the scientific breakthrough concerning EGF, and its ultimate value to the world, were thus shaped within a very particular geography of trust pertaining to Cuba: one in which warrant had to be secured on several different fronts simultaneously and where that work was factored into the scientific work itself. The question now was whether that "position" could be sustained.

SIX

# Strategic Marginality

At Marina Hemingway, a little west of the Science Pole, are the offices of Germán Rogés, York Medical's point man in Cuba. Like the business suites that have cropped up in the major Havana hotels, the marina is indicative of the new Cuba: a small step in a careful dance with capital. Rogés initially worked as a microbiologist at the Instituto Pedro Kourí, named after the notable Cuban physician and expert on tropical diseases. The Cuban equivalent of the US National Institutes of Health, the institute was founded in 1937 with a specific focus on parasitology and infectious disease research. At York Medical, however, Rogés now searches for financing for diseases that afflict the North as well; it seems the Cubans have reluctantly begun to accept that these pay better.

When I stop by Rogés' office one morning, he tells me that he sees the agreement between York Medical and the CIM as the most important outcome of the opening up of Cuban science in the 1990s. "YM has been a vehicle for scientists to be exposed to regulatory authorities, clinical trials, preclinical information, patent information, and so on [from elsewhere]," he says. He recalls some of the exchanges of people and ideas this has involved. "We took a whole series of people from some of the institutes here to Canada [but] it wasn't fully compatible and still isn't, though the situation has now improved. YM has proven that it can be worked on both sides: it doesn't just need to be management from outside. This is increasingly getting recognition from elsewhere. In the 1990s a few companies began to come here to check on the biotech situation. I would say beginning from 1991 you have Merck and SmithKline and the others interested. Awareness outside Cuba of our biotech here has become greater."[1]

## Bioeconomics

One such company was Germany-based Biognosis. Substantially more modest in size than Merck and SmithKline (now GlaxoSmithKline), Biognosis was a small outfit, conceived of and founded specifically on the idea of acquiring the rights to Cuban technology and sponsoring research in what it hoped was to be an emergent scientific economy.[2] In the mid 1990s Biognosis was one of the most active companies with an interest in Cuban biotech, after the Canadians. "I also spoke to representatives of ICI and Glaxo," CEO John Meers tells me when we meet up in Cambridge, England, of his attempts to establish a strategy with respect to the island. "But it's not mainstream activity for those companies—they had some more oddball people there keeping an eye on things, but that was about it."

Though when I spoke to him Meers was himself no longer confident about what is happening in Cuba, the Englishman was still keeping an eye on events on the island. He remains an inveterate Cuba watcher, waiting to see what might happen and, since joint ventures were legalized in 1992, occasionally testing the waters. "I've looked into about twenty projects in Cuba," he says. "One was a collection of crustacean shells to make chitin, as there is a market for this in various forms, mainly in slimming foods and in cosmetics. I tried to buy the bulk of it." The crucial issue, though—as we have been seeing—was how to access western markets: in this case the United States. "We had a system," Meers says. "But, the Cubans objected, they wanted to add the value and do it all *there*, they felt they were up to it, which probably they were. So they cut a deal with the Italians to market it as a highly specialized but tiny market product—we're talking kilograms—but with a very high value. The Italians said it would turn into a large market, but we didn't think it would, and it didn't."

Eventually Meers' company secured the backing of Beta Gran Caribe, the Cuban investment fund, to form a joint venture with Heber Biotec in order to market diagnostic equipment. The Cubans would supply novel reagents for use in diagnostic kits built by Biognosis. Reagents were a good way to go if the principal idea was to ensure access to first-world markets—of the various fields of application that biotechnology products might fall into, the market for reagents is a relatively immediate one (certainly as compared to the work on targeted drug therapies, say). But this time it was not Cuban reluctance so much as capitalist appetite that got in the way.

Beta Gran Caribe (later incorporated as Ceiba Finance Ltd) had investments across a range of sectors of the Cuban economy. It was taken over by the UK-based investment fund Laksey Partners, in a hostile move overseen by Laksey's Colin Kingsnorth, a "notorious investment-fund raider," as an article on the takeover in a financial magazine described him.[3] As for Beta Gran Caribe, "the whole thing is being wound up," Meers commented in an interview in 2003, along with his proposed deal with the Cubans: it wasn't profitable enough. Meers' story is not unique. From the mid 1990s there was a good deal of foreign investment capital tied up in funds earmarked for Cuba and eagerly awaiting the fall of either the embargo or Fidel Castro. But those funds sought profit not just productivity.[4] Meers goes on:

> The Cubans are unfamiliar with this and this is all the worst side of capitalism we are exporting to them. . . . The Cubans who come over to Germany are highly skilled and competent and relatively cheap. But there are lots of problems in getting them here. . . . I can understand why the Third World gets fed up with the First World ripping it off, and I go on marches with a different hat on, but it would have been dopey for us to [cut them an exclusive deal] with the chitin deal and it would have been silly for them because, in the long run, our project could have been back-integrated into Cuba and I would have been as good a friend as they'll find.

For Meers, the problem is one of trust all around, and the different experience in establishing this trust is what separates Rogés and York Medical's account of Cuban biotechnology from his own. "There are hidden barriers which are subtler than tariffs," as Meers says. It is the operation of such hidden barriers that require investigation: how is it that issues of trust, belief, and credibility, so crucial to the conduct and communication of science, might be bound up not only with investments but with the science being invested in? In Cuba, this in fact boiled down to a rather more specific question of how to manage forms of regulation and standardization—such as intellectual property rights—that are inherent to capitalist science. The global pharmaceutical industry requires these global standards and, even more, global coherence in the priorities that such standards are intended to codify. For Cuba, it was a question of how to determine what level of involvement in these mechanisms was "reasonable."

## Norming and Forming the Bioeconomy

Property rights exist, most basically, in order to demarcate the public from the private domain. The rules are no different when property rights are applied to intellectual property: here it is innovators who hold the private rights to own and sell their inventions.[5] Intellectual property thereby confers ownership rights to intangible goods, or ideas, which in turn become property in a legal sense.[6] The conferral of ownership to ideas and even to "invented" forms of life has lain at the heart of biotechnology's rapid development over the last two to three decades. And as the industry has grown, so has the system of intellectual property rights (IPR) been asserted globally. But the logic of intellectual property rights is a specifically Western logic. Not all societies practice property ownership in the same way. Certainly socialist Cuba historically has not, which in the 1990s presented Cuban scientists with something of a problem.

Information on how they got around this problem is hard to come by, but it seems that the basic trajectory of events was as follows. First, the Cubans found that in order to market their biotechnology to the world, they had no choice but to adhere to Western IPR norms. They sought to justify this—in light of the specific claims of that biotechnology we have been examining—by maintaining, at the national level, parts of the prior socialist system of intellectual property. Second, they were only able to do this because of the US embargo, which provided a sort of legislative loophole. Seen from the epistemic periphery, a more variegated form of the global biotechnology industry thus begins to transpire from what are otherwise simply practical solutions to the problem of doing biotechnology in the context of political-economic realities.

## Globalizing Intellectual Property

Patents are the most important form of IPR in biotechnology: they seek to marry short-term private gain with long-term public good by conferring on the proprietor an effective monopoly over the use of the patent for a limited period of time, normally up to twenty years. This, at least, is the argument that the developed countries (and the United States in particular) took to the Uruguay Round of the General Agreement on Tariffs and Trade (GATT) in 1986, where delegates addressed the question of what ought to be the international benchmark for intellectual property. They also took with them the growing complaints by US and European industries about the increase in pirated and counterfeited products coming from the South. These countries were concerned about the lack of any enforcement machinery within the existing intellectual property rights conventions (such as the Paris Treaty and even the World Intellectual Property Organization) to deal with the situation. They wanted a new enforcement mechanism and proposed what would become known as the Trade Related Aspects of Intellectual Property (TRIPS) agreement.[7]

Naturally, the developing countries were reluctant. They argued that the advanced industrial nations had been able to benefit from loose intellectual property laws when *they* were industrializing, and it was only fair to allow the South the same freedoms to innovate now. Intellectual property rights might be an incentive to innovation for the individual, they pointed out, but where the use of a protected invention is essential to the development of a particular technology in general, then patents actually act as barriers to more innovation at the social level in countries just starting out. As this disagreement deepened during the formation of the TRIPS agreement, Cuba was one of the Group of Twelve countries most active in tabling alternatives. None of these were successful, however, and TRIPS came into force as an international agreement to promote and secure intellectual property rights on January 1, 1995, to be administered by the GATT's successor, the World Trade Organization (WTO).[8]

The stated aim of the TRIPS agreement was to provide "comprehensive standards for the protection of intellectual property and the enforcement of intellectual property."[9] But in the small print, it was, as legal scholar Rosemary Coombe has argued, essentially a bargain between the developed and underdeveloped countries that offered better terms of trade to the latter if they acknowledged the interests of Western capital as being their own.[10] Effectively, it allowed the richer countries of the North to shore up their comparative advantage in intellectual and value-added goods.[11] This was not what the Cubans, now explicitly seeking some potentially innovative and marketable products from their biotechnology, wanted at all.

### Localizing Intellectual Property

Yolanda works at Lex SA, a Cuban intellectual property firm. In the 1980s, she recounts, the Cubans' approach to intellectual property hovered somewhere between the United States' and the Soviet Union's own understandings of the social functions and purposes of physical property. The Cuban approach to intellectual property had initially been formed very much in the mold established by the American Declaration of the Rights and Duties of Man, which speaks of "property" as something required by the individual to meet "the essential needs of decent living."[12] But when the Cuban National Office of Inventions and Trademarks (ONIITEM) was established in 1973, one of the first things it did was to declare the former US-influenced Law of Industrial Property of 1936 "a product and expression of the economic structure of its time" and one "which could not but reinforce the exploitation of our economic resources by foreign monopolies."[13]

ONIITEM immediately set about implementing an IPR policy that would resist the perceived exploitation of the former system. But it did not reject an American culture of ownership entirely. Unlike the Soviet Union, Cuba never rejected the notion of intellectual property outright.

The system it developed was in fact one in which innovation was construed as anything that might serve the national account, and thus came with obligations with respect to how those innovations were to be promoted or "worked."[14] This was set out in Decree Law No. 68 of 1983. Under this law, as is common in the assignment of intellectual property to researchers working in public institutions in the West, assignment of the right to work the invention (granted by means of a Certificate of Patent) was rendered distinct from the right to be recognized as the author of a particular invention (granted by means of a Certificate of Author of Invention).[15] But there were some important differences, however, on account of localized Cuban understandings of property and ownership. These, and how they relate to the patenting of biotechnology inventions, are rendered clear by reference to a document produced by the Cuban patent office the year before. This document clarified that a Certificate of Author was:

> the *fundamental* document for the protection of inventions in the socialist countries, through which is recognized the ownership [*paternidad*] of the author over the invention and the right to be remunerated according to the economic importance of the invention, while the exclusive rights to the use, exploitation, and commercialization of the invention is conceded to the state [*patria*].[16]

This is an important rider to Decree Law No. 68, and in particular Article 66 of that law, which allowed for the Certificate of Patent to be owned by the institute or research center in which the invention was made.[17] It essentially means that not only must individual inventors defer rights to work the patent to the institute in which they work, so too must those institutes in turn defer their rights to the state. This is a much stronger socialization of an individual invention than occurs in the West, and it conforms closely to the distinction between *paternidad* (ownership) and *patria* (homeland) in Cuba, which revolves in both cases around the affirmation of sovereignty expressed in the public rather than private interest. The preamble to Decree Law No. 68 specifically prepares the ground for this move: it splits the rights of inventors into "moral" and "material" rights. The right to work a patent was therefore ultimately written into *public* law as a part of the renationalization of the Cuban state and the concomitant reterritorialization of national sovereignty that was occurring through the process of Rectification discussed in chapter 3. The individual rights inherent in IPR were thus hardwired into Cuba's territorial sovereignty.[18]

This suggested interesting possibilities for Cuban IPR practice. But had things continued as they were, Cuba would likely have been gradually incorporated within the World Intellectual Property Organization (WIPO). Indeed, in 1989, just prior to the fall of the Soviet Union, there was a meeting of the Bureau of Inventions and Patents of the Council for Mutual Economic Assistance into which Cuba had been integrated since 1972.[19] Ten of the socialist countries as well as the WIPO were present. The final protocol of the meeting contemplated putting into action a program for the accelerated development of ONIITEM until 1995. Under this agreement Cuba would have received 20 to 25 percent of all patent publications in the world free of charge, in return for better acknowledgment of the rights of non-Cuban IPR holders.[20] But the proposed deal was eclipsed by the demise of the socialist bloc, and Cuba consequently followed a very different IPR route. After 1991 Cuba was left with a now vestigial system of socialist intellectual property coupled with limited bargaining power vis-à-vis such large Western institutions as the WIPO. Given the imperative to try and obtain hard currency from its biotechnology inventions, it needed to immediately find ways of actively integrating with that Western system of intellectual property.

## Order and Disjuncture

In many respects it was not therefore all that surprising when it was announced, on April 20, 1995, that Cuba had signed on to TRIPS.[21] As if to confirm its change of policy, the government additionally signed on to the Patent Cooperation Treaty (PCT) in 1996.[22] The result? Inventions that under Decree Law No. 68 had previously only been "protected" (i.e., ownership assigned) in a de jure and not de facto sense by the awarding of a Certificate of Author, would now have to be protected by patents under the replacement Decree Law No. 160.[23] And the text of Decree Law No. 160 duly acknowledged the need, at the highest levels of government, to fall into line with the international norms of a multilateral world.[24]

This appeared to many as a volte-face but it was played down by the minister for science and technology, Rosa Elena Simeón, as a "biotech rethink."[25] In a telling indication of the reasons behind the change, she later added that the main achievement of the year was to put science in line with the country's new socioeconomic priorities.[26] One thing that was certain was that the number of patent applications began to grow. In a discussion with Blanca Tormo, a representative of CIMAB—the CIM's business wing—in 2002, it soon became clear, however, that while now compliant with essential norms, the Cubans have retained also something of the former system. "[So] in general when you want to strike a deal on an alliance, or make a business development deal, [now] it has to have a patent, and the stronger the patent the better. It's [become] a necessary asset, a necessary tool, or item, but it's not the only one by any means: you have to have some other quality of uniqueness.... But I haven't said anything that's not in the book!" She laughs. "I mean I haven't told you anything that's not *there.*"

Indeed she hadn't. Article 3 of the new Decree Law No. 160 affirmed that any inventions realized in the course of a worker's employment by any state agency or institute (in effect, covering all work in biotechnology) would be submitted for a patent application "in the name of [said] entity."[27] State ownership of patents was still ultimately enforceable. But that "other quality of uniqueness" also interested me. As Fujimura has pointed out, meeting the requirements of the patenting process usually requires significant changes in the experimental and social work of the laboratory (and in the case of Cuba that might mean away from the model explored in chapter 4 to something else). With such wholesale changes involved if a country decides to go down the patenting route, it is understandable why patents have been described as an entire sociocultural paradigm, a "cultural infrastructure, which scientists need to understand thoroughly to be able to develop the technical and social skills needed for patenting." Intellectual property expert David Bainbridge notes, for example, that experience and careful drafting of claims are crucial in the patenting process, and here poor countries with little experience of Western legal norms are usually at a distinct disadvantage.[28]

So how can we explain the apparent willingness with which the Cubans took up the all too often stifling challenge of patents? Part of the answer lies in the felt need to commercialize the system and the possibilities that Cuba could *only* exploit if it were to develop an intellectual property portfolio. Also important was the need for the country's biotechnology centers to gain the acceptance of the international scientific community, and, of course, participation in international agreements, such as WIPO, garners a return flow of currency and technical assistance to signatory countries.[29] As one director put it—in a phrase to which we will return—"the rules of the game that we needed to play were changing." In some respects Cuba also had no choice but to sign up to TRIPS: by October 1997, 132 states had become members of TRIPS; it would clearly be difficult to opt out and arrange bilateral negotiations with each of them (something that China, working on a far greater scale, has been able to do successfully with the United States). All these were certainly incentives. But the most important answer is found somewhere entirely unexpected.

## The Space of the Nonspace

"There are many different difficulties, economic difficulties, living in Cuba," Luis Enrique Fernández, director of vaccine research at the CIM, reminded the readers of *Medicc Review*, in response to a question about how Cuba's scientific work has fared in light of the economic crisis of the 1990s. "But we are holding on to the hope that the only way to improve, to become a better country, is to be the 'owners' of the country ourselves. The moment we lose our sovereignty, we will lose this hope that we can be better."[30] Such hopes are revealing. On another visit to CIMAB, Blanca Tormo likewise talks of the embargo as a central part of the problem of trying to negotiate across two systems. "I mean you go to the negotiation table with disadvantages that other people don't have," she says. "The embargo is a huge disadvantage. It blocks out a very huge market, so that's one of the things which really [affects us]. But it's not something that plays at the forefront of your mind. You know it's there, you know it's going to be there, and you try to work around it." Time and again, in interviews and conversations, the embargo cropped up not as an explanation for the difficulties Cuban scientists face but as an element of the equation itself: as a constitutive part of their daily practice. How, then, did this artifact of the cold war shape the relationship between scientific work and forms of regulation in the present, and how, entirely against its stated objectives, would it offer the Cubans the means to work around the epistemological challenges to scientific practice that the problem of patents had raised?

The answer takes us back again to 1995 and the point when Cuba was signing on to TRIPS. At the same time, albeit for very different reasons, the United States was implementing a toughened policy with respect to Cuba. The collapse of the socialist bloc had already deepened the embargo's bite. Before 1989 the embargo affected the 15 percent of Cuba's international trade which fell outside the socialist market while from 1991 the embargo had a restrictive influence on more than 90 percent of that trade.[31] Then came Congressman Torricello's Cuban Democracy Act of 1992, which sought to restrict non-U.S. trade and credit relations with Cuba, and directly targeted exports to Cuba of products that might be used in its biotechnology program. But in order to seal the air lock further still, on March 12, 1996 President Clinton signed into law the Helms-Burton Act. Under Title 3 of this act, any companies deemed to be 'trafficking' in former United States property on the island would be subject to lawsuits by American nationals who were the former owners of that property.[32] The Cubans parried with various affirmations of national sovereignty, but it was a major blow to their hopes of obtaining increased foreign trade.[33]

It is obvious, therefore, that opinions on the embargo would strongly diverge either side of the Florida Straits. It is less obvious but equally pertinent that understandings of the embargo differ too. Even the dominant term used to describe it varies between the two countries: for the Americans, the word "embargo" connotes a technical-legal term: the act of sovereign rationality. For the Cubans, the word used is always *bloqueo* (literally, blockade), a political-moral term that sounds more like an act of filibuster. And directly in the crossfire of these legislative salvos were any companies that might wish to purchase Cuban biotechnology products or supply them with materials. Later on, the publishing of Cuban scientific articles themselves would be deemed by the United States Treasury to be in contravention of the embargo, a move resisted by the major scientific journals such as *Nature* and *Science*.[34] In such ways biotechnology became a stake in the most persistent struggle over Cuba's territorial sovereignty. In 1995, for example, Merck (America's largest pharmaceutical company) was fined and sanctions imposed when a number of its executives were caught bringing back biological samples of Cuba's hepatitis B vaccine from Havana for testing in the U.S. The fine in this case was $127,000, but fines for the purchase of Cuban products by Americans under the Helms-Burton Act can be up to $250,000 in individual cases as well as earning the accused up to ten years in prison. "Helms and Burton," as David Allan commented to me in 2003, some time after my visit to CIMAB, "must be simply falling over themselves with glee."

### *Sanctioning Knowledge*

For Cuban science, the tightening of the embargo during the Special Period was initially highly damaging. For one, it affected (and continues to affect) the supply of basic scientific materials such as reagents or gels or solvents. Seventy-five percent of the world's supply of insulin, for example, comes from neighboring Puerto Rico. Yet, since Puerto Rico is an American dependent territory, none of this gets to Cuba. One detects a certain bitterness, therefore, in López-Saura's observation that Puerto Rico "is nothing but a pill factory for the US." The expense of obtaining these inputs from elsewhere or via third parties increases their costs substantially.[35] "Go and ask Pérez what this means when he has to get reagents from third parties," says Salvador Moncada, a Honduran scientist and director of the Wolfson Institute for Biomedical Research at University College, London when I visit him in 2006. He earned his PhD for solving the mechanism of action for aspirin but is best known for his work on nitric oxide, for which he was the world's second most highly cited scientist from 1990 through 1997.[36] Moncada advised the Cubans on their biotechnology program back in the 1980s while Pérez was working on a temporary assignment at the Wolfson Institute. Pérez is back again in 2006 and I bump into him on the way out. I ask him about just this problem. In fact, he was reluctant to speak of it, but on record is the case of a Swedish corporation that was prevented from selling Cuba a sophisticated piece of medical equipment because it contained just one filter patented under US law.[37] The supply of raw materials is just one way the embargo impacts biotechnology science in Cuba, however. Even more important, the United States also represents around 50 percent of the market for pharmaceuticals: to have no access to either this or the socialist markets of the 1980s put a huge brake on the potential demand that Cuban biotechnology might have met and the corresponding income it might have realized.

But, with respect to innovation, the longer-term effects of the embargo are somewhat different. The relevant issue here is sovereignty. Despite both its seeming normality (as the bedrock of the international state system) and its normative effect, sovereignty is really just "an ideological representation of the way physical and social space are presently organised."[38] It is a fiction, that is to say. With some effort it is therefore malleable. By seeking to override Cuban sovereignty through its extraterritorial clauses, what the embargo actually achieves is to inscribe Cuba's particular form of property rights more strongly into Cuban territorial space. In so doing it perpetuates a space in which noncapitalist forms of property rights can be maintained in Cuba after the fall of the Soviet bloc, and it further removes that space from the foreign capital and its attendant forms of regulation that would fast make those socialist conceptions of property rights redundant. More specifically, the embargo as one form of hegemony (territorialized political hegemony) limits the purchase of TRIPS as another (deterritorialized economic hegemony). The embargo thus sustains

Cuba's alternative intellectual property system by virtue of "blockading" it from the influx of capital that is supposed to be the principal mechanism of TRIPS' own brand of "disciplinary neoliberalism."[39]

In a return visit to Heber Biotec, the CIGB's marketing wing and equivalent of CIMAB, I put some of these same questions to vice director Carlos Mella. Is it really possible that Cuba was, as this analysis suggests, quite literally in a position to maintain many of its own learned "sociotechnical competences" built up around a system of collective property rights? Mella responds, in a seemingly noncommittal way, that "the problem of patents manifests itself mainly in the First World." But if we set this alongside Rosa Elena Simeón's comment that signing on to TRIPS was not a volte-face, the two half-answers take on a more satisfying form, one which is most simply derived by rephrasing Mella's comment. It was not so much the case that the problem of patents was something that manifested itself mainly in the first world, therefore, so much as that the American embargo ensured that the problem of patents was unable to (fully) manifest itself in Cuba.

## *Atomizing Value*

This warrants consideration in more detail. Patents—that form of property that lies at the heart of the system TRIPS seeks to instantiate—quite literally exist as either paper or electronic documents. On to these documents are placed carefully redacted words describing a particular mechanism or process, through the sanction of which ownership of that mechanism or process is assigned. In such mundane ways, then, they code not simply for ownership but for a specifically Western idea of authorship *as* ownership. As critical legal scholars have pointed out—drawing on Foucault's analysis of the author-function—patents thus tend to operate as a sort of "global authorial empire": translating knowledge into economic value.[40] Knowledge thus becomes capitalized at the point patents connect it to economic forms of exchange. In the case at hand, however, the specific way that Cuban biotechnology knowledge was patented meant that ownership was codified in terms of a more local, socialist rationality, and attempts to maintain this in the face of the normalizing capitalist pressures of TRIPS constituted a struggle over knowledge between different systems of value.

To the extent that this epistemic struggle was mediated through property, however, it was also fungible with a larger political-economic struggle between the Cuban and American states over the issue of territoriality. This broad dispute concerns everything from the legitimacy of Cuba's socialist government to more concrete conflicts over the current ownership of once American-owned land expropriated by the Cuban government. This is a contemporary manifestation of a historical process, one described most precisely by legal scholar Kevin Burch when he refers to that fundamental "split in property (rights) [which] established the conceptual division between the state system (real tangible property—sovereignty) and the capitalist system (mobile, intangible property—money)."[41] "The institution of property rights," Burch continues, "contributes to the generation and linking of capitalism and the interstate system as articulated structures; differences between real and mobile property contribute to the differences between the two structures." But in the present context—in the realm of global biocapital—it is intellectual property rights, coding for a particular sort of value, which need to be considered in place of money as the mobile form of property.

The point, however, is that such mobile forms of property are cross-elastic with the more fixed forms of territorial sovereignty: hence, attempts to regulate Cuban political sovereignty came to have an impact on attempts to regulate Cuban notions of intellectual property. Moreover, such regulatory "overload" opens up a space in which the production of value becomes particularized, or atomized. Just how such particularization of value might be taking place can be understood in terms of the dialectical relationship between the logic of accumulation in global capitalism, on the one hand (expressed in mobile property rights—patents), and the logic of state sovereignty, on the other (expressed in fixed property rights—territoriality). The principal expressions of these two logics of power and regulation as they relate to Cuban biotechnology are TRIPS and the Torricello Helms-Burton embargo. Since TRIPS and the embargo are two different ways of regulating the same space, the overlapping of these two logics of power results in a case of "regulatory overstretch" in which the dialectical relationship between the two logics collapses.[42] In this case, the principal forms of mobile property rights concerning Cuban biotechnology (patents) are thereby given a geographical fixity. And in turn, the fixing of the economic value that patents code for allows the epistemic (or authorial) value that they also code for to be rather more free-floating.

The consequence of all this is that the nature of the value of a Cuban patent varies according to where it is taken up: within Cuba or without. Without such freedom of movement it is hard to see how value can be subject to anything but universalizing processes (such as the attempt to globalize intellectual property norms). With it, however, the Cubans managed to maintain a space of operations for their particular biotechnology practice by securing their more local definition of value against the global trend. The Cubans could thus continue to develop drugs that are not economically viable elsewhere but which were socially valued in Cuba: epistemic contiguity with the past had been maintained by virtue of political-economic disjuncture in the present.

As a result of this territorial loophole, Cuba has to some extent been able to continue to formulate its intellectual property policy both within international IPR structures, such as the WIPO, and on a case-by-case basis whenever this suits it more. As a director at the National Patents Office put it, "Our system now has the best of both worlds." Indeed, as of the early 2000s Cuba had 150 patents registered in Cuba, 66 in other countries, and 500 applications pending. Not all of these were biotechnology patents, although the meningitis B vaccine alone was patented in nineteen countries as of 2006. But at the same time, Cuba continues its practice of selling its open-sourced technology to those countries not accepting US patents: of the fifteen countries Cuba has technology transfer agreements with (Argentina, Brazil, Canada, China, Egypt, India, Iran, Mexico, Malaysia, Russia, Tunisia, South Africa, Venezuela, the United Kingdom, and the United States), many have themselves sought to resist the uptake of US patents as a global standard. In short, outside of the US market (admittedly a large share of the most lucrative Western markets), the Cubans are to some extent able to operate both inside and outside of the TRIPS framework in a way that legal scholar R. Weissmann and others have shown was not possible for other developing countries.[43]

## Critical Effectivity

While I am at Heber, Carlos Mella asks me to review a promotional video for the CIGB with him. It finishes with the catchphrase for Heber Biotec hanging over a horizon: "All the way from the idea to the product."[44] It's an attempt to capture the idea of full-cycle research, he explains to me— or, given that Heber is the CIGB's marketing wing, to extend it geographically, to other markets perhaps. CIMAB and Heber are not the only market-oriented institutions in the Science Pole. As part of the commercialization drive considered in the previous chapter, each of the main biotech centers of the Science Pole got its own dedicated marketing office. They act as bridging points between the rules of the global economy and the logic of Cuban science. As Mella explains, "Heber functions as a commercial mechanism so that the CIGB [can] continue its [research and production] work." He goes on: "The achievement of commercial relations is not contradictory to the fact that we are a socialist company [*empresa*]. We are a little more creative with what we do with the income generated. The idea of course is that we work for the social benefit, and so the profits go to a central state fund which is then distributed throughout the Science Pole according to the need of the various centers."[45] That is what makes the video so important. It is one element in another broader strategy of boundary work. But what I wanted to know was what other adaptations, aside from patents, had taken place *within* the Science Pole?

A few immediate developments can be readily pointed to. Most obviously, the Scientific Council set out to internally reform the working practice of the Science Pole in the way described by José de la Fuente earlier. Biotechnology, the Scientific Council believed, could be commercialized without compromising its evident capacities for innovative work. To put this into effect, the five-year plans of the 1980s were replaced by yearly plans which themselves were "revised on a monthly basis at the meetings of the Scientific Council," Eduardo Silva, the director of Biomundi, tells me. But as one of the marketing representatives repeated over and again with respect to patents, buying into a different system of value isn't the only thing. It's a different approach to science that you are talking about if you want to make capital a part of it. As we saw in the previous chapter, such considerations shape how you define the nature of the scientific problems you plan to address. And so it was that from the formation of new regulatory policies within the centers, focused on issues such as "biosecurity" (meaning laboratory safety and quality control, not the current Western use of the term) and a growing concern with the level of "professionalism" to revaluing what was meant by "value" itself, substantial changes were afoot in the very conduct of Cuban science. There is no accessible record of these events in Cuba. What the following analysis shows, by drawing on interview material to try and recover something of that history, is that while these changes were more positively embraced than the responses to the normalizing effects of IPR, such processes of formalization would in fact do *more* to undermine the innovative mode of operations that had developed in Cuban biotech in the 1980s.

### *Instituting Rationalities*

As the date at which CECMED was founded suggests, such moves toward formalization were already taking place before the fall of the Soviet Union, but in the 1990s these two bodies together formed the two central pillars of a new National System of Regulation and Sanitary Control.[46] With the establishment of this system the production of domestic medicines was no longer the core priority. Emphasis was now placed on strengthening the regulation and control of that production in order to secure foreign sales of Cuban pharmaceuticals. CECMED was intended to "evaluate, register, inspect and analytically verify medicines and diagnostics,"[47] with the aim being to ensure that a fully independent body integrated and carried out all regulation of pharmaceutical products.[48] It therefore displaced regulatory functions from the state organs previously responsible and formalized the whole process by setting clear protocols in place of the case-by-case approach that had been dominant until then.

CENCEC, on the other hand, was created in order to improve the clinical trials system in Cuba. The experiences of attempting to market the meningitis B vaccine and other products had drawn to the Cubans' attention the economic importance of ensuring that all evidentiary data pertaining to its drugs was conducted not only in a rigorous manner, but with a form of rigor readily understandable by and acceptable to foreign entities such as EMEA and the US Food and Drug Administration (FDA). This does not mean that Cuban science before was dangerously unregulated. It means simply that the way in which it was regulated was not in line with Western norms. CENCEC initially focused on prioritized products so as to facilitate their rapid registration and subsequent marketing abroad.

CENCEC operates as the Cuban equivalent of a Contract Research Organization (CRO)—undertaking the management of the whole clinical trials process, from selection of patients to analysis and review of data.[49] It does not charge the Cuban centers for its services, but it does charge foreign firms. All of this demonstrates a certain continuity with the previous model of mutual assistance. So too does the way it is funded: a fixed grant from the state in national money to pay the workers, and a second sum, also from the government but in US dollars to purchase equipment, reagents, and so on. As the director of CENCEC openly commented: "We can't do it all on our own, so we have to invent a little from time to time," to which she added, "we are currently working with a Bahamian company registered in a third country to avoid the embargo." While there are important differences between these two institutions, one thing that can be said about both of them is that they have been flexible and adjusted their approach to the changing requirements of the field, including expanding their operations, cooperating with other institutions, developing new areas of expertise, and so on. To some extent they also act as interfaces between the institutes of the Science Pole and international regulatory regimes such as TRIPS, and therefore they also internalize the rationality of organizational decentralization consequent to those regimes.⁵⁰

In this they are not alone. In 1991 the Cuban Office of Inventions and Trademarks (ONIITEM) was completely reconfigured as the Cuban Office of Intellectual Property (*Oficina Cubana de Propiedad Intelectual*, or OCPI).⁵¹ While still nestled protectively under the Ministry for Science, Technology, and the Environment, the OCPI was intended to exhibit a greater independence from the center in order to facilitate the adoption of IPRs. ONIITEM, it was now suggested, had been hindered by disorganization, incomplete information from the socialist countries, and a lack of integration with Western countries and their intellectual property offices in particular. Just as in the 1980s the scientific institutes had been left alone to learn the practice of biotechnology science, so in the 1990s was the OCPI left alone to learn the equally important practice of biotechnology regulation. Under its control were five former state offices, which were now reconstituted as independent processing offices for various sorts of IPR claims.⁵² The research institutes are now unable to publish results without the approval of these offices in order to ensure that the work has not contravened existing research that may already be subject to patents: which, after ten years of non-recognition, the Cuban institutes are at a greater risk of doing than many. Hence, while patents remained legally embedded in the state, the state nevertheless affords these offices the freedom to implement that policy in a way that best fits with international requirements. Meanwhile, intellectual property specialists were put into all the centers (sometimes researchers were taken out of the laboratory and given retraining in this field) in order to ensure that the new protocols were indeed set in place.

In contrast to all of this, other institutional developments of the 1990s brought a greater degree of political recentralization to the Science Pole. The Scientific Council (with de facto authority over the Science Pole now that it controlled the Biological Front), for example, was co-opted into the newly empowered Academy of Sciences (ACC) in a clear example of political recentralization. The ACC operated as a conduit for state policy and its proposals were generally approved by the State Council (the highest administrative authority in Cuba). The directors of the most important scientific institutes were also, at different points, co-opted into the political process. Luis Herrera and Rosa Elena Simeón became members of the State Council (the thirty-one-person governing body of Cuba), Concepción Campa joined the Politburo, and José Miyar Barrueco remained secretary of the Council of State, further centralizing the decision-making process.

On a per capita basis biotechnology arguably became the industrial sector with the greatest political insertion.[53]

Biomundi also came to play an important role in the reorientation of Cuban biotechnology toward international norms. Through the publication of its yearly directory, bilingual technical glossaries, and guides to potential investors, along with a yearly "Who's Who in Cuban Biosciences" publication, Biomundi operated as what Bruno Latour would call a center of calculation for Cuban biotechnology science, compiling and making available information from elsewhere: drawing the research institutes into a more coherent whole. In so doing it intended to reduce the very boundaries that had separated Cuban biotechnology from the global biosciences to that point. Biomundi even offered tours of the Science Pole and the famed beaches of Varadero for a "complete business package." But in categorizing and defining in minute detail the functioning of the Science Pole in terms of who worked on what and where, who to contact about what, and which equipment was to be found where, it rendered the different institutes and research groups visible to foreign parties through their work on specific areas only.[54] Something of the earlier willingness to resist specialization had been lost.

### An Architecture of Reason

It was through the above forms of institutionalization that global scientific rationalities truly became factored into local Cuban reason. The same took place as Cuban modes of organization were then additionally deterritorialized: the logic behind them being given over to non-local entities. The ISO-9000 series, for example, is a set of norms and commonly agreed upon protocols that seeks to ensure quality control in production, including biopharmaceutical production.[55] It is the basis for systems of certification at the international level and for many countries is seen as a key to international markets.[56] The International Standards Organization (ISO) which implements these protocols has been in operation since the 1960s to reduce differences between national regulatory systems and has been a fundamental principle of the WHO since 1969. The standards fulfill a valuable role but, like TRIPS, ISO norms are also part of an international regulatory architecture that is often culturally girded by hidden Western norms. ISO standards then map those norms into a wide range of areas, from mobile phones to paper sizes to textiles. Quality management standards, for example, which instill such priorities as a "factual approach to decision making" are becoming increasingly important if companies want to obtain preferred supplier status. The norms are periodically revised to take into account technological and other developments and have increasingly come to be taken up within public sector and governmental domains. This is done on the basis of feedback from existing users, the most dominant of which are large organizations whose own particular logic of management structure thereby becomes incorporated in the "international" norm. ISOs are thus voluntary but skewed and pervasive. Since the ISO was founded in 1947, some 16,000 standards have been published. In every case, local (Western) reason is converted into global rationality. But how does such reason touch back down elsewhere?

In April 1991, Cuba joined what were then around fifty countries that already utilized the ISO-9000 norms and approved its version of them—the NC-ISO 9000 series—along with a revision of existing concepts regarding quality control in Cuba.[57] This created an immediate problem in that implementation of these norms has varied between the centers. While the Finlay Institute, for example, has applied for ISO certification, the CIGB has not, pointing instead to its Good Production Practices certification awarded independently by the WHO in 2001. The result has been a certain divergence of policy within the Science Pole. But more profoundly it has resulted in the instantiation of a new approach. Put simply, quality was now "the principal objective of every organization"; before it had been *an* objective.[58]

Quality control is of course a major factor in biotechnology production: products must be subject to national and international norms, certification, metrological assurances, and registration. Some experiments can only work if they run in air-locked rooms inside other sealed rooms (a "box in a box" as some describe it). As suggested above, Cuba had much to do in this area to meet international standards but it took rapid steps to catch up. Good Laboratory Practices, for example, were first implemented in Cuba in 1992, and in 1994 minimum standards of product management (storage, transport, etc.) were implemented for clinical trials.[59] Over the next five years, a series of further minimum standards were imposed regarding collection of biomaterials and medicines, information provided for the public about medicines produced for the national market, and studies of bioavailability and bioequivalence.[60] These standards were then themselves expanded in 2000, particularly with respect to clinical trials.

The point here is that such "valuations" of work are culturally constructed in the same way as is intellectual property. And that means that in order to comply with them, various things had to change in the Science Pole. Even its very infrastructure had to be brought into line with the new protocols. In many labs this meant that the HVAC system to control heating, ventilation, and air conditioning of critical support was installed. This was important anyway for monoclonal antibody laboratories, which require a high degree of atmospheric purity, but it also allowed the Cubans to satisfy international legal and regulatory norms, providing monitoring facilities to give supporting evidence of the standards being met and the costs being reduced. Such developments signaled a growing conformity with international protocols in biotechnology. But so too did they belie a more significant change in the Cubans' thinking. We encountered earlier the Cubans' relaxed approach to scientific risk taking, and the importance of this to their overall results. The issue of risk was the principal theme taken up in a Cuban doctoral thesis of 1996 which noted that a model of biosecurity "had not previously existed in the country": hence a lack of attention to clean surfaces, no emergency plans, and most tellingly, no system of reporting incidents and exposures.[61] From the mid 1990s this all changed, with authority deferred to the regulatory capacities of bodies and institutions operating from outside Cuba at the same time as those bodies' norms were factored into local scientific practice.

What had emerged most strongly of all, then, was a concern to *demonstrate* full quality control mechanisms.[62] In a 1995 study into the possible design and application of quality control systems in the institutes of the Science Pole—a report that used the word "deterritorializations"—L. Gómez-Napier, a titular researcher at the CIGB, argued that, "Now it is no longer sufficient to obtain and maintain quality, but rather it is necessary to create confidence and demonstrate the existence of a system of the required quality."[63] The point, it seems, was by then being well heeded. The introduction throughout the Science Pole of these various international norms we have been examining was part of a growing recognition of the importance of securing credibility and warrant for Cuban scientific practice.[64] In their approach to patents, the Cubans may have rejected the imperative of capital as a means of valuing scientific work, but in making their own work fungible with the capitalist system—that is, in achieving a critical level of effectivity—a simultaneous reterritorialization of some of the guiding rationalities of the Science Pole and a deterritorialization of others had taken place. In the process the very spatial organization of the Science Pole—that crucial element to Cuba's erstwhile experimental mileu—had changed.[65]

## Liquid Science

The global biopharmaceutical industry is fast moving. I earlier characterized Cuba's fledgling biotech industry of the early 1980s in terms of the speed with which it was able to reverse engineer interferon. But as we began to see in the previous chapter, as the global economy was increasingly factored into local science, it was not so much speed as a more generalized state of "liquidity" that became the defining feature of Cuban biotechnology science: crystallized in the problem of how to merge. A former Cuban scientist, Jorge, captures these problems of epistemic and political-economic intersection illuminatively. "The country was in a big crisis," he recalls from where he is based now in a lab in Wisconsin,

> and they were trying to change the economy a bit. They were trying to decentralize it you know. This was happening in biotechnology as well . . . it was clear now where the money was. We had money or we didn't. But the money was very slow moving, you know. They could say you have seven thousand or eight thousand dollars, or ten or twenty—this was around the money they had for projects. But that doesn't mean you can use it tomorrow. You have to wait until the money is liquid. That wasn't the exact translation they used to say, but anyway, the money is probably credit in some other form. But by the time the money is liquid it could be at the end of the year, so you have to buy everything the following time. Otherwise you have to a wait a lot before all the ordering gets processed. So at the end it's not so efficient.

José de la Fuente makes a similar point in describing the fate of scientific labor under such conditions: "This field moves so fast. If you take a scientist out of his lab and keep him out for like six months, without access to the journals and so on, then he is completely lost."[66]

Against these turbulent developments, it is a valid critique that socialist science proves too rigid; that the same reasons a drug is made available to all become the same reasons that drug is not subject to further refinement and development outside of the original setting in which it was created, and why its capacity may not, therefore, be maximized. But the point that I want to take from these two comments is not as to the efficiency or otherwise of Cuban science, so much as for what they tell us about the logic under which local (and primarily Western) reason is scaled up into global rationalities and then touched down into local contexts, just as we saw with the cultural norms of ISO standards. We learn from Paul Rabinow that such norms are only ever partially assimilated into actual social forms. But what are the implications of this for relations of power in pharmaceutical science?

One possible answer, as we have seen in this chapter, is that attempts to regulate science, to formalize it, can be resisted in unanticipated ways.

The web of global norms that tries to give particular shape to the pharmaceutical industry may well be a machine for the regulation of expertise, but overregulation of liquidity can have unexpected effects. The American government's political regulation of Cuban scientific activity contributed to a sort of regulatory loophole—the space of the nonspace—in which Cuba's existing scientific practices could to some degree continue to take place. If such liquidity seems to proliferate beyond the scope of any simple pro forma—if it is too liquid—it is not without a certain strategic logic, however. And this, as I have tried to show, is because of the way it becomes invested in cultural systems and norms. The failure to understand this explains why the Cubans' attempts to capitalize on their biotechnology failed, at least initially: they simply did not notice that they were in the process reconfiguring the very architecture of reason on which their earlier successes had been based.

## Strategic Marginality

What this also shows is that, like any other regime of truth, the structure of a Western rationality of science articulates a particular sort of will to power focused on the determination and sanction of reason. Like any other regime of truth, it is also therefore bounded, and it is at the margins that such systems emerge in their purest form: at the point where their power relations are resisted only to reinstantiate new power relations. That is what we see with the dual regulation of Cuba as a specifically disenfranchised site within the US-dominated geography of global pharmaceutical production: one that is therefore, ironically, most capable of resistance. Here is where value is revalued. After all of this work, the new "value" of Cuban products was that they could now be differently conceived of in different places; they could be a part of one system of value (global intellectual property) at the same time as being part of another (Cuban socialist intellectual property). This was something novel indeed, and it reminds us that we can only ever "know" (and so produce) value from particular political-economic positionalities. While many epistemic terrains might therefore be most readily colonized by global capital, others are more easily taken advantage of by marginal players. These players may have no choice but to play the game but, as Wittgenstein reminds us, the real art of engagement is in having a stake in the rules themselves. This is how resistance may be formed.

Such resistance is in part about the production of countertruths. "That's the other issue that is important," Blanca Tormo comments as we discuss the recent (much encouraged) rise in Cuban publications. She continues,

I think that [publishing] really was a window that opened on an issue that made everybody look to Cuba, not only because they realized it was biotech and it was good science and we had these products and intellectual property and so forth, but because it opened a window of possibility of negotiating with the United States. I mean [through these products] was the first time there was a retraction of the embargo to be able to negotiate.

But, as we have seen, resistance is not simply about the production of countertruths; so too is it about acting in relation to those truths, in this case, by making marginality into a strategic asset. "Both CECMED and CENCEC follow international guidelines of the ICH [International Conference on Harmonization], EU and so on," York Medical's point man, Germán Róges, informs me. "Despite the political differences, representatives of all these have come to Cuba at some stage. They will try to harmonize as much as possible but each country will be slightly different." This isn't just known, accepted. It is actively worked on. Slippage is a necessary feature of regulation. Even simply the incorporation of joint ventures outside of Cuba is a standard approach now to the problem of the embargo. This is how, as Ian Hacking might say, the market has been bent toward Cuban science.

But markets are unpredictable. My conversation with Blanca Tormo took place shortly after the recent BIO event in Toronto, and it wasn't long before we ended up talking about recent events at ImClone—the biotechnology industry's scandal du jour. That year, ImClone's CEO Sam Waksal had been arrested the day after the BIO event in what Blanca refers to as the "Watergate of biotechnology." Waksal was later indicted before the Southern District Court of New York for defrauding ImClone by selling off his own shares in the company (and advising family and friends, including American home-design guru Martha Stewart, to do the same) after learning that the FDA had refused to approve a Biologics Licensing Application for Erbitux, its anticancer drug (an experimental monoclonal antibody not unlike Cuba's own).[67] Such a licensing application would have been the first step to gaining regulatory approval and to ultimately selling the drug. "Sam was here in 2000 with a group of other people and he didn't seem such a greedy man!" Blanca jokes, "He's really put a dark spot on the biotech industry in general, because, well, ImClone *is* biotechnology and a fraud like this puts a black spot on the whole industry because it decreases credibility, in general. It's not an isolated event, because it makes the whole industry . . . people start to think, well how can I be sure the rest of biotech is trustworthy?"

Issues of trust and confidence may not be more important in biotechnology than in other domains of scientific research, but they do pose a particular problem for non-Western operations. Indeed, if the Cubans are particularly sensitive to such issues, it is for the simple reason that they know full well that political-economic burdens of proof fall disproportionately on them. That much was made acutely clear in 2002 when, after nearly a decade of struggling to find the appropriate means of reengaging in a global marketplace, the United States Undersecretary of State John Bolton accused Cuba's biotechnology industry of producing biological weapons.

SEVEN

# Peripheral Assent

The tensions that arose in the early 1990s between the "solution" to the problem of normalization and the "problem" of formalization that the solution posed were dealt with in a straightforwardly political fashion at the close of 1998. In December of that year Manuel Limonta and José de la Fuente were just the most visible scientists removed from their posts in what de la Fuente describes as a crackdown within the institutes of the Science Pole. Limonta had played a central role in the development of Cuban biotechnology. Castro had needed someone like him to make biotechnology work in the first place. He had also needed Limonta and other entrepreneurially minded directors in the early 1990s when Cuban biotechnology was first reengaging with the world. Limonta had excelled in this task too: *Science* labeled him an "entrepreneur who had made Cuba's biotech investment pay dividends."[1] Before his removal was announced, Limonta's office was searched by officials, with the reason given that he had "reached the limits of his capacity to develop the center," the removal of certain individuals being more generally described as a "restructuring of personnel."[2] Whatever the truth of that matter may be, it marked a return to the more bureaucratic approach to science of the 1970s.

And yet, on the surface of it, things were looking good. In April of the same year, Sergio Pérez, the general manager of Heber Biotec, had announced on the basis of the Cubans' actual and planned joint ventures with such companies as Biognosis that the CIGB-Heber enterprise "ranks twentieth worldwide from the standpoint of its earnings and around eighth in net profits."[3] This was enticing news for the UK business mission organized by Britain's Caribbean Trade Advisory Group (CARITAG) and Cuba's Foreign Trade Ministry, which was to come to Havana later in the month to discuss biotechnology collaboration opportunities. "The point is," Agustín Lage declared at that event, "can we open a channel to Europe through the UK?" Such a move had long been championed on the other side by British trade minister Brian Wilson, and there was considerable goodwill and interest on behalf of a good number of British organizations.[4] But Lage also lamented that the big pharmas had thus far been slow in approaching the island. "Cuba needs strategic alliances to balance our upstream strength against our downstream weaknesses."[5] Cuban biotechnology, he says, has been like a "well-kept secret" that the world should now be in on.

That particular offer was reproduced time and again, from the *Financial Times* to *Time* magazine, and there were signs that some people might be keen to take it up, British pharmaceutical giant SmithKline in particular. Of all the big pharmas, SmithKline is the most heavily involved in infectious diseases (which account for around a third of its current product pipeline). Its directors were well aware of the lack of an effective meningitis B vaccine and that preliminary indications in the mid 1990s were that the Cuban vaccine was based on "good science." But, while a British outfit, SmithKline's R&D takes place in Belgium at a partly American-owned subsidiary. Before they could do anything with the Cuban drug they needed approval from the US Treasury and, in particular, the Office of Foreign Assets Control (OFAC). SmithKline submitted a request for a license to develop the Cuban drug in 1998. Its success depended on a rather different interpretation of what constitutes "good science." In a revealing slip, State Department spokesman James Foley commented to a press briefing when asked about the grounds on which the application would be considered: "We are aware of the appeal, *of the proposal, rather*, by the company to work with Cuban researchers to develop this vaccine. I can only say that appropriate United States government agencies are discussing the issue."[6]

To help them do so, SmithKline corralled a range of expert testimony that, interestingly, had on the whole little to do with the science. Admittedly, in North America, Carl Frasch, the expert on meningitis B who had aided the Cubans (their approach was closely modeled on his work) and who was now chief of the FDA Bacterial Polysaccharides Laboratory, wrote in support of SmithKline's development of the Cuban drug on scientific grounds, stating that one to two thousand cases of meningitis B could be prevented in the United States each year if the Cuban vaccine were properly developed and marketed. But so too, on rather more political grounds, did sixteen House members write to Secretary of State Madeleine Albright asking the State Department to grant the license. In July of 1999, after two years of trying, SmithKline was finally exempted from the provisions of the US embargo and granted a license to conduct clinical trials for the US market.[7] Under the terms of the deal, SmithKline was required to pay the Cubans first in food and medicines and then ultimately in royalties when the drug reached market.[8] The torturous legal requirements notwithstanding, the SmithKline deal was potentially the biggest contract yet for the Cubans. "It's given a lot of confidence to other US companies that might have been less confident before," Allan said, when I spoke to him again—though likely with a certain trepidation at the intrusion of the big players onto what had until then largely been "his" island.[9] At the same time, another major global player, Amersham, developed an interest in Cuba and secured $2 million in sales of materials to the island, mainly to Servicex Departamento 4, a Cuban company which sources supplies not otherwise available in Cuba for much of the Science Pole. In 1999, the president of the international division of Amersham was quoted in *Cuba Business* as predicting that in coming years, Cuba would become "the regional centre for medical research and development."[10] Cuban science had just been quite literally put on the map.

## Back at the CIM/EGF Again

Also in 1999, the US Patent and Trademark Office granted one of the CIM's drugs, TheraCIM, a patent. The European Patent Office would do the same in April 2000. TheraCIM had been developed alongside the CIM's more radical work with EGF and was by now its lead drug. In October, San Diego business leaders—who form a powerful regional coterie of biotech interests—announced an interest in establishing an informal relationship with Cuba and specifically the CIM. Like so many others, they wanted to be in a working relationship with the Cubans if the embargo ended. The Chinese already were: building on the previous agreements to sell diagnostic equipment produced at the CIM in China, the CIM now signed a more extensive deal with China's Ministry of Science and Technology. For a while it seemed as if confidence was growing and that finally some marketing results were beginning to be seen. Official discourse on the island promoted a positive image. Regular mention again began to be made of "mak[ing] the impossible possible."[11] In an article entitled "The Pharmaceutical Industry: A Bet for the Future," Cuban scientist Ernesto Grillo noted, for example, that to be able to meet national need with homegrown medicines may seem "science fiction" to many, but in Cuba "it is possible." As if to support his claim, BIOCEN, a Cuban biologicals production center, was granted ISO-9002 certification, fully endorsing its products from several of the biotechnology centers. All these were seen as indications that biotechnology in Cuba was doing ever better.

By March of 2000, on the basis of the previous years' successes, Peter Scott, the chairman of soon-to-be-dismantled Beta Gran Caribe, publicly anticipated real increases in its biotechnology investments in three to four years' time.[12] Although this would turn out not to be, it captured the optimism of the moment. Indeed, with the collaboration between the CIM and York Medical going well, the main concern was that Cuba's biotechnology sector required a more selective use of capital. As part of a rationalization process initiated to effect this, a prospective study aimed to identify new areas for Cuban biotech to focus on, and associations with foreign capital remained top of the agenda at the meetings of the Scientific Council.

Cuban-American politics from January to March of 2000 were completely defined by the Elían Gonzales affair. But despite this Castro visited South Africa and signed a series of deals with South African President Thabo Mbeki. One of these cleared the way for the two countries to cooperate in producing low-cost AIDS drugs. The process would involve ignoring international IPR protocols unless the South African government were to win its landmark court battle with thirty-nine large pharmaceutical companies to facilitate access to cheap medicine (which it did: the case was later thrown out of court). The nonaligned movement was supportive of South Africa's case, and speculation was rife at this time that middle-income countries—Cuba, South Africa, Brazil, India, and Thailand in particular—could collaborate in the manufacture of cheap drugs and pose a serious challenge to large Western drug companies. For now, however, this really was no more than science fiction. The most promising part of the sector in Cuba remained the CIM–York Medical collaboration, and that, in turn, rested on their front-running drug, TheraCIM.

TheraCIM was the culmination to date of the work put into cancer therapeutics in the early to mid 1990s. It is a humanized monoclonal antibody which directly targets the epidermal growth factor receptor (EGFR) for the treatment of epithelial-derived malignancies. Along with the so-called small molecules class of drug that such antibodies compete with, these antibodies are tyrosine kinase inhibitors, a category of drug that had emerged as one of the most important targets for cancer therapeutics.[13] It is thus more like other Western immunotherapeutic approaches than the vaccine—it is a passive approach. If approved, TheraCIM was set to be the first humanized monoclonal antibody approved specifically for head and neck tumors, and only the third approval in this new class of cancer drug for solid tumors overall.[14] Head and neck cancers caused nearly a quarter of a million deaths in 2001, while nearer to half a million new cases were diagnosed.[15] With antibody technology more generally expected to constitute a third of all biotechnology drugs over the next decade, TheraCIM promised to be the major, fully realized breakthrough of Cuban biotechnology to date.

## The Burdens of Proof

In July of 2000, the *Journal of the National Cancer Institute* ran an in-depth article on biotechnology in Cuba, focused on the CIM–York Medical partnership. It marked the beginning of significant news interest in the project. The majority of such articles were characterized by outdated statistics and a highly optimistic accounting of projects in the pipeline, but Allan doubtless appreciated the media coverage as his company expanded and diversified. In December 2000, York Medical changed its name to YM Biosciences, Inc. (YMB). Its product portfolio became much broader, and at least half of this came through alliances that YMB formed with non-Cuban companies, notably with smaller British and Canadian biotechs, and in part this was because it proved much easier to form alliances elsewhere. As Allan said, "The first thing people want to know is if we are breaking any laws by dealing with them [the Cubans]. Of course we are not breaking any laws but there is this mindset that won't go away."

Perhaps this was why the Cubans also developed a renewed focus on other collaborations. Agustín Lage made a one-week trip to China where he inaugurated the offices of the Beijing-based Biotech Pharmaceutical Ltd Corporation, the joint venture signed in 1999 between the CIM and China's Life Center (a subsidiary of the China International Science Center of the Ministry of Science and Technology of the People's Republic of China).[16] Cuba was showing signs of being able to make its own way in the world after all. As if to confirm this, in October the UK-based protein-engineering company Biovation announced a license agreement with the CIM that granted Biovation a nonexclusive worldwide license to the CIM's antibody engineering technologies. Frank Carr, CEO of Biovation, stated that "research work at [the] CIM in monoclonal antibody therapies and vaccination is world class."[17] In December, Missouri Senator Ashcroft, the man who helped push through European acceptance of Monsanto's Roundup Ready Cigars (Monsanto is located within his constituency and Ashcroft is effectively an international ambassador for the St. Louis firm), was also pushing for Monsanto to follow in SmithKline's footsteps.[18] The big pharma interest Lage wanted appeared to be in the offing—and not from Europe after all. Monsanto had been eying the Cuban market for years, it said, and it sent four representatives to the Biotechnology Havana conference that November to sound things out further. Monsanto wanted to capitalize on Cuba's access to third world markets: it offered a backdoor policy in return for the rights to Cuba's own proprietary inventions. Again, Cuban biotech was being positioned geographically, but this time not simply by being put on the map of global pharma. Now it was potentially being offered a strategic role to play.

### *Lage's Article*

At this point, the Cubans made an attempt to further clarify the IPR situation in Cuba. Agustín Lage published an article in the Cuban journal *Biotecnología Aplicada* entitled "Biotechnologies and the New Economy: The Value of Intangible Assets."[19] While not so widely read, Lage's article was important. The article rehearsed many of the issues that are characteristic of Cuban biotechnology but added a novel twist. Lage presented the situation in global biotech as one in which economies of scale and new forms of knowledge privatization internalize knowledge within capital, eroding the already scarce competitive advantage of poor nations. In addition, Lage understood the consolidation of biotechnology in Cuba as an example of a resistance (*enfrentamiento*) to the Western model of privatized biotechnology. As set out in his article, he sees this resistance as based on certain underlying factors, most notably:

1  The design of centers of research production. This is the groundwork for basic research being undertaken within a framework of applied research.
2  Complete cycle operation. This provides flexibility.
3  State guidance. This ensures research is directed toward the public good.
4  Social property. State ownership of property in *conjunction* with decentralized management is a productive one.
5  Export orientation. This complements state funding.
6  The integrated focus of Cuban biotechnology within a multi-institutional system. This allows integration and cooperation between centers.

Interestingly, his article begins with the question of what Cuban biotechnology is: "The most important first step is to ask, who are we?"[20] He adds that the lack of distinction between research and production within Cuban centers blurs the divide between these aspects and so problematizes the former question. But so too does the growing influence of companies such as YMB and the fact that it is *they* who are also determining how and why Cuban biotechnology is now developing and where, as a result, it remains successful. The implications of this were made clear by an IPR expert at the Cuban firm Lex, SA.

> The CIM have been very good at developing IP, they are the most developed center in the Science Pole in this regard and are already receiving money from licenses abroad based on patents: this is more than the other centers have achieved. We have tried to be a bit strategic with our use of patents, using "old friends" when necessary because patents were designed for the capitalist world and we need to "translate" them a bit.[21]

Who, then, was involved in this sort of translational work, and how were they seeking to represent Cuban biotechnology such that the discourse, if not the practice, of science might more easily "carry" the sort of slippage described in the previous chapter? As I have been arguing throughout, representations are not merely ethereal attachments: they help to organize and construct the world as we see it. As John Law puts it: "representations shape, influence and participate in ordering practices: that [is to say] ordering is not possible without representation."[22]

Throughout the 1990s, Cuba's biotechnology effort was organized under three dominant discursive frameworks: natural resources, the family, and the national corporation. Each sought to combine what would be readily digestible means of describing science with a Cuban flavor. The natural resources trope, for example, appears with considerable frequency in Cuban and foreign investment literature. In 1994, in the first edition of the magazine *Avances Medicos de Cuba*, targeted at foreign readers, the country was described as "in the process of development but in full scientific bloom."[23] David Allan, for one, had picked up on this. "We also used the natural resources metaphor a lot, i.e. the idea that if you have copper in the ground you take it out when it's valuable, not before or after. It's about the present value. We spent a lot of time with them on the idea of present value, because that's what matters in this industry."

Alongside this, Cuban scientific endeavors were often understood in terms of what we might call "familial" science. This set of representations has consistently been used to describe a family of scientists, oriented toward assuring healthy families. It is part of the preventive health discourse in Cuba, of course, but it also plays out at the business level. The early advertising material produced by the Cubans played especially on this trope. An advertisement for Finlay's set of vaccines, VA-MENGOC-BC, Vox-SPIRAL, VA-DIFTET, and Vax-TET, for example, appeared in various places, including the annual Havana trade fairs and magazines such as *Avances Medicos de Cuba*, with the slogan: "Behind every healthy family . . . there is a family of vaccines."[24] Within these descriptions the biotechnology centers themselves became seen as a family of institutes—we have heard this slogan before, in interviews—with CENIC as the "mother" organization. Here then are various attempts to relocate the imagination of Cuban biotech to a more intimate, less threatening scale in such a way as to reconnect the current political-economic realities with the traditional precepts of Cuban health.

Finally, and as a logical extension of the metaphor of Cuba SA we encountered earlier, the idea of a national corporation appears in various places, such as when the director of neurosciences in the Science Pole describes the institutes' "spin-off" companies, such as Heber or his own Nueronica, as "minitransnationals."[25] In these descriptions, Cuban biotech has become "safely" located within the language and formulations of global business. The above tropes thus contribute to a sort of representational landscape of Cuban biotechnology that exists in relation to its material referent in no less functional a way than does fictitious capital relate to material wealth.

For the Cubans, engagement in such branding is in part the unavoidable consequence of having to market their products. But it is also a response to broader suspicion of their work.[26] Indeed, such representations as shape and define Cuban biotechnology are not just produced within Cuba. As the above examples suggest, such images circulate and, as they do so, they take on new meanings. And in 2002 came a discursive shift that would develop into a far more visible struggle to redefine Cuban biotechnology than the relationship between YMB and the CIM or SmithKline's dispute with the US treasury. It was a way of representing Cuban biotechnology that was heavily overdetermined by geographical and in particular geopolitical imaginations. It was the point, in fact, at which the limits of representational convergence exemplified in the notion of a Cuba SA were not only reached but breached.

## *The Cancer of War*

Throughout the 1980s and 1990s, Castro had also described biotechnology using a variety of military metaphors: as a form of social "weapon" or a line of defense in the "battle" for public health. In this he was quite in tune with the discursive realities of Western medicine, where the idea of "targeted" therapies and "magic bullets" also weighed heavily on both public and scientific imaginations. At the opening of the CIM, for example, he described how monoclonal antibodies were like "guided missiles" targeting only the problem cells and not the good healthy "citizenry" of the body.[27] Other uses can be found in the references to "cadres" (a term for military groups) of scientists, an "arsenal" of Cuban products, and so on. All of these tropes are drawn from a broader discursive field: Castro's proud emphasis on Cuba as a world medical power as a way to define global status based on medical, not political, intervention. (Cuba's medical internationalism is legendary, and it is an often repeated fact that Cuba has regularly had more medics stationed abroad than all the G8 countries combined, for example.) But this most positive of tropes was reconfigured in the new millennium as the United States became concerned that Cuba was exporting not only biomedical technology but also the technology with which to produce biological weapons. Far from a world medical power, therefore, Cuba was in this representation a haven for weapons of mass destruction. So it was, then, that the attempts to form international alliances, such as between YMB and the CIM, became tied into these broader political and rhetorical disputes over biotechnology.

## Reasonable Doubt

The skies were already darkening over the biotechnology project in Cuba when I first arrived in Havana in late 2001. A political storm was soon to break. First, an article by the former but now exiled vice director of the CIGB, José de la Fuente, was published in October 2001, condemning the CIGB and Cuban biotechnology in general as having failed Cuba's original "model" biotechnology. Then, eight months later, in May 2002, even more damaging claims were made as then US Undersecretary of State John Bolton alleged Cuba's involvement in biological weapons production and technology transfer to "hostile states." For the media it was a free-for-all (though Castro wasn't slow to capitalize on the spotlight either) in which the careful work that had gone into forming a bridge between Cuban science and various quarters of the global biopharmaceutical industry was overlooked in the haste to trumpet the "clash" of cultures between the diligent West and the rogue elements of non-Western science. For the scholar these events need to be set in their proper context, however, as part of the more generalized crisis precipitated by the Special Period and the island's subsequent attempt to export its biotechnology without relinquishing what

Cuban scientists held to be the basic tenets of socialist science. That is to say, we need to tell this story once more from the Cuban perspective. This requires reading the crisis of 2001–2 as the culmination of the events set in train in the early 1990s, a crisis that shows the continued rescaling of the internal/external dialectic described at the close of the previous chapter. Indeed, both moments of crisis were rooted in the same geopolitical tensions inherent to the paradox that is global science.

### *"Wine into Vinegar"*

In October of 2001, however, it was political rather than economic burdens of proof that took center stage with the publication of José de la Fuente's article in the highly respected publication *Nature Biotechnology*. Since his dismissal from the CIGB, de la Fuente had left the country and was now at Oklahoma University. By the time his article "Wine into Vinegar" was published, phase 1 clinical trials of the CIM-YM merger's lead product, TheraCIM hR3, had just been successfully completed and phase 2 trials in Canada were being initiated by YMB. The terrain on which these trials were to take place now threatened to change.

> For myself and many young Cuban scientists, the establishment of an advanced biotechnology centre in Havana was the most challenging and rewarding endeavour we had ever undertaken. When our dreams were realised, the Cuban biotechnology program was a landmark in scientific accomplishment and a source of pride in the developing world. Less than ten years later that vision has been shattered, betrayed by a combination of intrigue, infighting, and bureaucracy. Cuba's once-vigorous biotechnology is now on the verge of expiration, strangled by increasing social and political tensions. Lacking capacity, creativity, and credibility, it is a paled and perhaps dangerous shadow of its former self."[28]

So reads the introduction to de la Fuente's article. While partly a polemic against his former colleagues, it was also an erudite piece of scholarship, offering a well-crafted and lucid history of biotechnology on the island and what he saw as the "political crusade" launched against scientists who "diverged from the hard party line" and were thereby branded as politically aberrant. "In the nineties they [the government] were losing power and possibilities, so they started again to enforce those political considerations, which resulted in a political crusade against the scientists. . . . This was new in the scientific sector," he told me in an interview. For de la Fuente, biotechnology in Cuba had become in the 1990s the white elephant many had (wrongly) predicted it would become in the 1980s; so far as he saw it, the reason for this was the excessive meddling in the affairs of scientists by a state concerned only with regaining complete political control over their conduct. For de la Fuente, it *was* a case of Lysenko redux.[29]

## Havana Responds

The response from Havana was a strongly worded letter to the editor of *Nature* by Dr. Julián Alvarez Blanco, president of CIREN, dated October 19. Alvarez Blanco lambasted de la Fuente as a "biotechnology Judas." "Evidently, De la Fuente is looking to curry political favour to justify his scientific failures and desertions, searching for new and less demanding horizons as to his personal conduct and moral values."[30] Neither so eloquent as de la Fuente's piece nor as well pitched a critique, it was, revealingly, a response penned in the vocabulary of the past. He then went on to state that, despite the difficulties of the Special Period that may have led some scientific workers to have to ride to work, Cuban biotechnology had continued to maintain a "rhythm" of good results. My aim here is not to determine the veracity of de la Fuente's allegations of a "political crusade." There is, in any case, a more useful task to attend to, and that is to look at the discursive field within which these claims were produced and to consider how they impacted the broader cartographies of trust against which the Cubans' drugs had to secure credibility. What values do such claims appeal to, contest, or seek to reshape? What issues are considered to be important? These became more clearly articulated in two further critical moments.

## Back at the CIGB: Biotechnology Havana 2001

In November of 2001, shortly after my own arrival in Cuba, I went by taxi out to the CIGB for the first time. The driver who took me was well aware there was a conference in the *Polo* that week; I was not the first person he had taken there. The Biotechnology Havana conference takes place over one week in November every other year. It is how the Cubans keep tabs on the world. It is the major event on the biotechnology calendar in Cuba, a chance to represent the work being done there to a select audience of "friends" and new visitors (including myself at this point) and, just as important, a chance to establish commercial agreements with foreign companies. Yet while the conference undeniably had an upbeat air, it nevertheless articulated certain elements of the more general crisis de la Fuente's article described. First was the concern for (and lack of) longer-term credit. Announcements that deals were in the offing with Iran did little to assuage fears that the sector was in need of significant investment. The foreign delegates expressed this more vocally than the Cubans. The crisis of the early 1990s—that is, the dire need for investment capital—was far from being unequivocally put to rest. The "doors are open for business," Lage claimed, but as Allan pointed out, this didn't meant that just anyone would be let through them. Second, bureaucracy is everywhere in evidence. It takes considerable effort just to secure a number of interviews with Cuban officials, something that involves an extensive vetting procedure. While waiting I spend a lot of time talking to many Cuban and foreign delegates, from a wide range of the centers. There is an uncanny similarity among their accounts of the current work in the Science Pole, suggesting a party line that, as I would later discover, was notably tighter than in previous years. Third are the figures, poorly publicized from the official statistics, for the export of medicines and scientific equipment, which declined from just over 47 million pesos in 1997 to around 33 million pesos in 2000.[31]

### Bolton's Charge, May 2002

The second phase of the crisis unfolded against this backdrop. On May 6, 2002, John Bolton, US undersecretary of state, alleged "Cuba has at least a limited offensive biological weapons program and might be transferring its advances to other states hostile toward the United States."[32] Bolton's claims appeared to draw loosely on de la Fuente's article and on the claims of the former deputy director of the Soviet biological weapons program, Ken Alibek, who, since his defection from the Soviet Union, had become a regular contributor to policy debates on biological weapons.[33]

They also drew on a comment Castro allegedly made during a speech at the United States University in Tehran. Castro was supposed to have said, "Iran and Cuba, in cooperation with one another, can bring America to its knees." It seems that Bolton did not check his facts, however, for as research has subsequently shown, the elusive quote has no provenance.[34] Quite simply, Castro never said it. Bolton's claims did have a clear target, however—the Science Pole, and the production facilities at the CIGB in particular—and they occasioned considerable international interest.[35] The response by the Cuban state was decisive. It immediately galvanized the most prominent institutions of the Science Pole into producing a series of statements denouncing the claims. On May 11, Castro himself, flanked by Roberto Robaina (former minister of foreign relations) and the various directors of the Science Pole, Herrera, Lage, and Negrín prominent among them, made a public response, printed in full in *Granma* and broadcast on national television. "The only thing true in Bolton's lies," Castro countered, "is the geographical fact that Cuba is situated ninety miles from the coast of America."[36] He took the opportunity to remind the world of suspected incidents of biological terrorism against Cuba by the United States.[37]

Castro also took the opportunity to list, in full, the achievements of Cuban biotechnology.[38] On the same day as the broadcast, a *Tribuna Abierta*—a mass demonstration of national support—was held in the district of Regla in eastern Havana to protest against the claims and affirm the integrity and national importance of the Cuban biotechnology project. Around 120,000 people were reported to have attended.[39] A special edition of *Granma* was printed with photos of the directors amid a sea of Cuban flags; their well-crafted speeches were reprinted in full.

### Carter's Visit, May 13

In a further twist, these events also coincided with a visit to the island by former United States president Jimmy Carter. Carter, who had been the only United States president to consider reinstating diplomatic relations with Cuba, was sympathetic to the island's biomedical endeavors.[40] His trip was immediately reworked around a visit to the Science Pole, a tour of the CIGB, and then a subsequent ceremony held at the Latin American School of Medicine, all broadcast on state television and watched by a Cuban audience rapt with the idea of a United States president on Cuban soil.

### Crisis Resolved

Back in the United States, Colin Powell moved quickly to tone down Bolton's claims. While the United States believed Cuba had the capacity to develop biological weapons, he said, it did not have evidence that it was actually doing so.[41] Representative Stephen Lynch, a Boston Democrat who had visited Cuba and who sat on a subcommittee on national security, said the Bush administration's attempt to portray Cuba as a supplier of bioterrorism materials was aimed mainly at influencing an upcoming vote in

Congress on whether or not to ease the embargo. "It's just politics," he said, rather unsatisfactorily.[42] De la Fuente himself clarified that Cuba was not producing biological weapons. Rather, he said, the concern was in its transfer of *potentially* dual-use technology to countries on the US watch list, Iran in particular. By May 13 the United States had retracted its allegations. *Granma* offered its verdict: the Cubans had it right. If the Americans were going to dominate the world system, they should expect that countries have to resist in order to make their own way forward. Albert Sasson spoke out in Cuba's defense.[43] Carter left. The debate subsided. Life returned to normal in the Science Pole.

The level of certainty a juror must have to find a defendant guilty of a crime is defined as needing to be beyond "reasonable doubt. . . . a real doubt, based upon reason and common sense after careful and impartial consideration of all the evidence, or lack of evidence, in a case."[44] But science of course works not with certainty but with uncertainty and, as discussed in the introduction, any doubt is only ever "reasonable" within a particular cultural (or even political) context. Just as in previous chapters we saw that the truth-producing mechanisms of science must themselves be produced in the world, so can we now also see something of the wider cartographies of trust and credibility against which these determinations are made. They are different sides of the same coin of a reason presaged on doubt. As Colin Powell said, the United States did not actually have evidence. What it had was the possibility inherent within uncertainty, something Powell's successor would push to the extreme of "unknown unknowns."

## Peripheral Assent

When I speak to them today, the Cubans are understandably frustrated that their industry is portrayed as having so much baggage. It is perhaps all the more frustrating for them that the one question most consistently asked is whether, after all of the burdens of proof they must overcome, the Cubans' drugs actually work the same as anyone else's. As should by now be clear, the only possible response to this question is to reject the terms on which it is asked. Anthropologist Andrew Lakoff reaches a similar conclusion in his study of what he terms "pharmaceutical reason." Lakoff examines psychiatric practice in Argentina, set against the development of diagnostic norms in Europe and North America. He explores the problem that ensues for North American drugs when the illnesses they were developed for turn out not to "exist" (or at least, not to have the same diagnostic value) in Argentina: how then to make an illness to fit the drug?[45]

The Cuban case presents similar fundamentals. Cuba's drugs must find their way in a global pharmaceutical market in which they too are "out of place": the dynamics are the similar, and at stake is the same issue of how a scientific product is valued differently under different regimes of truth. But the specifics are somewhat different: here it is how to make drugs like Cuba's work in other markets. It is a question of insertion. To work in the world, Cuban drugs must gain two competing sets of confidences: first, of the Cuban state, which insists that they be made according to distinctly local principles; and, second, of the regulatory institutions of the WTO, of bureaucrats at the Office of Foreign Assets Control, of the scientists operating under strict protocol at the FDA and so on, which demand they adhere to international norms. They must gain these confidences as much as they must convince scientifically. One consequence of this has been the way that companies such as YMB have used discursive strategies—the idea of Cuba SA, for example—to create a sort of middle ground, to gain the confidence of competing regimes of truth. A more pervasive consequence, however, has been the gradual appearance of a new entity: a distinctively socialist drug.

It is impossible to make a "socialist" drug, of course, but socialist drugs are what Cuba's drugs have become, and both the Cuban government and the key stakeholders in global biotechnology have made them so. Like it or not, the Cuban products *are* a result of their context in this way; they bear not just the mark of Heber or of CIMAB, but the marque of a particular epistemic milieu, with all the layers of meaning which that implies and which we have been uncovering. In the 1980s the Cubans somewhat explicitly argued that their products had a socialist component to them, though in fact they were really the product of various practical efforts to get around the rigidities of socialist science. In the 1990s, by contrast, the Cubans sought to demonstrate the opposite: that their drugs are just like any other, that there is nothing special at all about the particular context in which they are produced. The position of the Cubans' work relative to global biotechnology, however, has gradually made a reality of what was wrongly claimed in the 1980s and wrongly denied in the 1990s. To the extent that they are a product of the way that their development in a socialist country has been taken up within the capitalist world economy, Cuba's drugs today really are socialist drugs, the product of what they are perceived to be, and the impact of those perceptions back on the practical elements of drug production: methodology, clinical and preclinical testing, evaluation, regulation, and marketing.

Douglas Star, codirector of the Center for Science and Medical Journalism at Boston University, reports on an interview with Carlos Borroto of the CIGB that highlights this:

> Borroto . . . remembers talking to colleagues about using patents to protect their expanding market. That was the moment Castro decided to pop into the lab. "What's all this about patents? You're sounding crazy!" he said. "We don't like patents, remember?"
>
> Borroto stood his ground. "Even if you're *giving* medicine to the third world," he said, "you still need to protect yourself."
>
> Borroto knew he had to get better at the game. He sent his staff to Canada to get MBAs, to learn the language of capitalism. Yet concepts like venture capital still escape him. "I can't understand how 80 percent of the biotech companies in the world make money without selling any products," he says. "How do they do this? *Hopeness*," he guesses, using a neologism to stress the absurdity. "They sell *hopeness*."
>
> Asked for an annual report—a basic necessity of international business—Agustín Lage, director of the Center for Molecular Immunology, merely says, "You know, we've actually been meaning to produce one." Then he smiles and shrugs.
>
> It's like Castro said: They don't really *like* patents. They like medicine. Cuba's drug pipeline is most interesting for what it lacks: grand-slam moneymakers, cures for baldness or impotence or wrinkles. It's all cancer therapies, AIDS medications, and vaccines against tropical diseases.[46]

The "hopeness" they lack—and it is actually not such a bad phrase—is speculative capital. In reality, this is all that they lack. But so central to the myriad ways that modern pharmaceuticals get valued has capital become that its absence (admittedly, a diminishing absence in the case of Cuba) requires the void to be filled. The production of Cuba's drugs as socialist drugs is one way of doing that, though it results in a new realm of problematization: socialist drugs in a capitalist world economy. As to what then becomes of such socialist drugs, two examples are particularly revealing. One is a deal struck with the Indian biotech company Biocon in 2002; the other, a deal struck with San Francisco–based CancerVax in 2004.

### *Biocon: Electronics City, Bangalore*

"Genomics has converged biologicals and pharmaceuticals, defining a new and exciting interface *biopharmaceuticals*," announces Biocon CMD, Kiran

Mazumdar-Shaw on the Biocon homepage. "At Biocon, we leverage our unique skills and resources to capture the biopharmaceutical space." It is illuminating that the CMD of one of India's oldest biotech companies (founded just two years after Genentech, in 1978) should choose the idea of "capturing space." What sort of space was she referring to? The global market, or a particular subset of it? Or the space of biopharmaceutical knowledge itself? A brief historical excursus helps to clarify. For much of its past, Biocon was focused on enzyme synthesis. As it set about reinventing itself as a drug discovery company, it evidently found that it needed a new corporate vocabulary to go with it. In that sense, it was of course much like the CIM. Like the CIM also, and as pointed out by the anthropologist Kaushik Sunder Rajan, who has studied Biocon in detail, Biocon's real asset was its manufacturing capacity. "[It is not] that U.S. biotech companies do not scale up their manufacturing as they expand—it is just to say that manufacturing scale-up is not the aspect of their business that gets play[ed] as part of corporate PR and investor relations, as the activity fundamentally driving valuation. . . . [A] call to build a company through the validation of its manufacturing capability, is virtually the opposite of Genentech's [an exemplary American-style institution] history."[47]

In February of 2002 CIMAB and Biocon signed a joint venture agreement creating Biocon Biopharmaceuticals Private Ltd (BBPL) in which Biocon held a 51 percent stake with paid-up capital of Rs 8.8 million (about $0.2 million dollars in today's money). Like CIM, Biocon's "integrated business approach" covers everything from early stage drug discovery to clinical development and marketing, and the Bay area companies have a history of sharing intellectual property. As Mazumdar-Shaw puts it: "I firmly believe that much of the work in biotechnology has to be in collaboration because you need both a good strategy and sophisticated infrastructure."[48] Both structurally and ethically that puts Biocon alongside the Cubans' own "full-cycle" organizational model, but what is of most interest is that this South-South agreement is centered on precisely that most materially valuable of commodities (as opposed to the speculative value of many North biotech entities): not "hopeness," to borrow Borroto's phrase, but a manufacturing system proportionate to need. Thus, out of this alliance comes not just a business plan centered on different ways of sharing intellectual property rights but a 120,000 square-meter, cGMP-compliant multiproduct biologics production facility in Bangalore designed to manufacture an Indian version of the Cubans' EGFR vaccine using the Cubans' own out-licensed technology. It is a large facility (the largest in India to date, in fact) for a large market. In 2001, world figures for head and neck cancer were estimated at 454,446 new cases. Biocon has the rights to market the Cuban drug in India.

### CancerVax: Carlsbad, California

The Biocon agreement took on an additional interest when in July of 2004, CIMAB also gained a unique license from the US Senate to out-license three further experimental cancer drugs to California-based CancerVax Corporation. The drugs are the specific active immunotherapeutic candidates previously licensed to YM as part of the original suite of cancer products they had taken on. While YM retained the rights to TheraCIM, it now out-licensed three other drugs (Her-1, the lead product, targeting epidermal growth factor; TGFa, targeting transforming growth-factor alpha; and EGF-ECD, targeting the extracellular domain of EGFR) to CancerVax, a newly public company which at that time had no products on the market. The deal covered both clinical development and manufacturing of the vaccines. For this, CancerVax made an upfront payment to the CIM of $6 million payable over three years, with an additional $35 million and royalties payable to CIMAB on commercialization. While YM received various licensing and milestone payments and retained an interest in the revenue from any eventual marketing of the drugs, it had no further obligations with respect to them.[49] CancerVax got the rights to market the drugs in the United States, Europe, and Japan, and Biocon got the rights to India. "This is a great endorsement for CIMAB's cutting edge science," Mazumdar-Shaw said at the time. Little mention was made by the US government, or the media, as to whether this undermined the recent claims of biological weapons production, however.

In fact, approval for the deal was given by the US Office of Foreign Assets Control (OFAC) in an explicit exception to the US embargo. But in order for the deal to go ahead, payments to Cuba during the developmental phase were required to be in goods like food and medical supplies. Even to secure this level of approval, considerable work was required. Two separate law firms were employed to make the case. Ultimately, however, according to an anonymous spokesperson for the State Department, the reason for the license was the drugs' potential "life-saving" ability. In a comment in the *New York Times*, the spokesperson said: "Upon review it was decided that the company should have an opportunity to further research and verify the claims about these drugs."[50] Little mention was passed of the various oncology experts who were drafted to provide expert testimony, again both confirming the importance of the production of "veracity" that Steven Shapin describes and reminding us how such veracity is produced outside of science or, rather, at its own margins. As if to further confirm it, CancerVax's own literature points out that "in addition [to the possibility of the OFAC license being revoked] we cannot be sure that the FDA, EMEA or other regulatory authorities will accept data from the clinical trials of these product candidates that were conducted in Cuba as the basis for our applications to conduct additional trials, or as part of our application to seek marketing authorizations for such product candidates."[51] The agreements with the North are substantially more fraught, then, than those with the South, as the Cubans' earlier move into the Chinese market in part recognized.

## "Here's to Us, Who's Like Us?"[52]

So how did things finish? In June 2003, YMB floated on the TSX and London's AIM, raising $15 million in the nick of time for the company. While this represented the bottom end of the financing the company was hoping for, it was nevertheless a victory of sorts in an otherwise heavily depressed biotech sector. YMB's was the first successful biotechnology IPO in Canada that year and one of only three in Europe. At the same time, in Cuba, the Science Pole began to be called the Western Havana Bio-Cluster. A subtle shift in terminology perhaps, but an indicative one no less. From 1999 the CIGB was working in collaboration with the Center for Science and Technology Management to introduce strategic planning as a central part of the center's rationale, one intended to rekindle certain of the former elements, particularly those of "motivation" and "leadership."[53] The success of the CIM, it seems, had by then been noted. By the time of a meeting at BIO 2002, a new face of Cuban biotechnology seemed to have emerged.

While the Cubans remained hesitant toward capital, by 2003 they were building manufacturing capability in both India and Iran. In retrospect, therefore, Lage's article ended on a prescient note: "Once again, what has been done is exactly the opposite to that which is proposed for the vulnerable scientific systems of Latin America by the neoliberal recipe. But this has been precisely the task: denounce reality and construct an alternative."[54] It remains to be seen, however, if such an alternative can be fully realized and maintained. Senator Christopher Dodd, Democrat for Connecticut, wrote to Secretary of State Powell in 2004 urging that permission to license Cuban drugs in America be granted on medical grounds.[55] "Saving lives shouldn't be a political issue," he wrote, in a direct echo of Randolph

Lee Clark's colleague before the Texans' visit to Cuba back in 1981. But, of course, that is precisely what it is.

And what of the Cubans' edge in that heated EGFR market? Oncology firms, patient groups, investors, and regulatory bodies alike are all watching closely, and not only because the potential market for these drugs is huge. So too do they represent the first real competition for targeted therapies as an entire class of drug. Things came to something of a head at the American Society of Clinical Oncologists meeting in June 2004, where the now five EGFR inhibitors squared up against each other. If more than one drug reaches the market for the same indication then it will come down to an issue of cost, and, as Allan pointed out, manufacturing capacity could prove decisive then.[56]

In India, it already has. In 2006 it is the Cuban-developed drug subsequently licensed to Biocon that inaugurated India's largest biotechnology region: Biocon Park. As Indian president A. P. J. Abdul Kalam noted, in dedicating the drug to the nation, "Although a large number of monoclonal antibodies have been introduced into the country by multinational pharmaceutical companies, they are beyond the reach of the majority of cancer patients due to their prohibitive cost. The present indigenous development of monoclonal antibodies will be accessible and affordable to a larger number of Indian patients. Biocon joins the exclusive league of monoclonal antibody developers and will become a key player in this segment in the coming years. I would like to congratulate the researchers and developers."[57] To add glamour to political heft, Bollywood star Shahrukh Kahn unveiled the product, the first anti-EGFR humanized monoclonal antibody for cancer to be made available anywhere in the world. According to MarketResearch, the global monoclonal antibody market reached sales of $14 billion in 2005. It was expected to exceed $28 billion by 2010. And the antibody was Biocon's first proprietary drug for a market in which the Indian subcontinent represents one-third of the global burden of head and neck tumors. But the trace of Cuba had been almost entirely removed: its products confirmed once more as socialist in the moment that the threat of such a possibility is strategically avoided.

Just a few months later in 2006, Lage published another position piece, this time on the problem of cancer immunotherapy. The answer is not to look forward, he counsels, extending the logic of his work to date, but instead to look back at the historical connections between immunology and cancer research, to go back to immunology's preventive roots. This goes against the cancer industry's generally top-down approach, as historian of medicine Roy Porter notes: "in the cancer industry, . . . the philosophy of prevention has never been very successful in attracting big money and top people—it goes against the grain of our favored high-tech, commando-style interventionist medicine."[58] But Lage's comment makes perfect sense from the Cuban point of view. Indeed, if we are moving to a technoscientific society characterized by the sort of "vital politics" the sociologist Nikolas Rose describes (that is, a politics centered upon our genetics and our biology and that calls for the management of and intervention in personalised risk landscapes), then there is a certain irony in the way that it may therefore fall to places like Cuba to meet the equally growing pharmaceutical needs of more vulnerable populations.[59]

## A Global Grammar

As the previous chapter suggested, many people involved in biotechnology around the world have been looking to events in Cuba for some time. This chapter has set out to explore what it was that some of them thought they saw and to consider the implications for Cuba of the various attempts to define its biotechnology (whether discursively or through the practice of biopharmaceutical partnering) on the continued development of that science or, rather, on the efforts to maintain a space for that science amid a rather differently couched global pharmaceutical norm.

On the surface the claims made by John Bolton appear somewhat antithetical to those made by José de la Fuente. While de la Fuente's claims were based on the perceived failure of Cuban biotechnology, Bolton's were based on its perceived success, albeit one of aberrance and alleged danger. For others still, such as Ken Alibek, these two sets of claims are not antithetical but part of the same broader problem: the failure of Cuban biotechnology to undergo a limited commercialization while retaining what had made it successful in the 1980s forced it into selling its technology to hostile states in a last-gasp grab for cash. Where the danger lies, here, is in the unknown waters of South-South technology transfers. But for those such as the CEOs at CancerVax and Biocon, this was precisely what made Cuba such an attractive proposition for an alliance.

The brief surfacing of these various views in the same politico-scientific space (the space of the Cuban biotechnological) reveals the centrality of trust to the operation of contemporary biopharmaceutical science, and how different cartographies of trust impact a place like Cuba. Modern life, and the forms of science it supports, would be impossible without what Niklas Luhmann call "system trust." As Shapin argues, "Trust is no longer bestowed on familiar individuals; it is accorded to institutions and abstract capacities thought to reside in institutions."[60] So too, might we add, is it more easily ascribed to certain of these modern intermediaries in some places than in others. Indeed, it is this latter character of trust that allows system trust to exist: system trust requires boundaries to be drawn in the world: complex things like science work because they are done in a particular way, the "right" way, often an unacknowledged "our" way. Trust may be a global grammar, but it is one with specific, local effects.

EIGHT

# The Cuban Cure

Recognising the connections between ideology and science should prevent us from reducing the history of science to a featureless landscape, a map without relief.

—Georges Canguilhem

In a speech entitled "Science Matters" given to London's Royal Society in February 2002, British prime minister Tony Blair spoke of a "fundamental distinction" that he saw "between a process where science tells us the facts and we make a judgement; and a process where *a priori* judgements effectively constrain scientific research." "So let us know the facts," he added, "then make the judgement as to how we use or act on them." It was an issue which animal rights protests and the controversy over genetically modified food in the United Kingdom had given the prime minister strong feelings about. What made him finally decide to speak out, he said, was a recent visit to Bangalore—where Biocon Park is based—the month before his speech. Researchers there had told the prime minister that Europe had gone "soft" on science and that it stood to be leapfrogged as a result. Blair saw his government's hands-off support of science as evidence to the contrary. But his glaring positivism—assured of the separateness of science from any form of politics or cultural values—rather missed the point. Or rather, it mis*located* science with respect to politics.[1] Rather than to draw overt comparisons between Western and non-Western forms of science, or to try to tally the pros and cons of strong state support for science versus a more laissez-faire approach, what I have tried to do in this book is to rather more carefully locate Cuban biotechnology science; to consider its imbrication within the Cuban state and localized forms of rationality set against broader global forms of reason instantiated in the neoliberalized global pharmaceutical economy. I have sought in the process a history of the relations through which this Cuban work and some of its principal outputs were constituted.

## Pharmaceutical Orders

When they first embarked on their scientific journey into biotechnology science, Cuban scientists were given a very explicit set of orders: to put together a functioning biotechnology industry on the epistemic periphery. Immediately, then, did the two worlds of science and political economy collide in a rather more explicit way than usual. Theirs was to be a specifically socialist science, they were told; it was to confront head-on the health problems of the day, and it was to be nurtured by the state. The particular scientific order that emerged, however, was more contingent than this: not just socialist but nationalist also; not just about overcoming adversity but about bringing together a novel set of interests in order to do this; not just state-supported but girded by the informal supports that developed within that outwardly formalized milieu. What emerged was an experimental milieu for the conduct of biotechnology science and one that proved highly efficacious. In the process, one thing soon became clear: that what was (and continues to be) taken as a quintessentially Western and capitalized form of science was here the product of a rather more localized scientific rationality. In short, there were (and are) other ways of thinking and doing biotechnology.

In my early chapters, I considered what this other way of doing biotechnology looked like and, in particular, how a novel epistemic milieu was put together in Cuba on the basis of a certain epistemic distance from Euro-American dominated biotechnology science; that is to say, how a Cuban form of scientific reason was predicated on certain locally sustained norms of political and economic rationality. Later I considered how this structure was problematized when Cuba's experimental milieu was forcibly connected to a broader global rationality in which value was determined less by social utility and more by economic efficiency. This was not simply a case of the disembedding of local reason in light of forces of globalization, however. In fact a series of far more subtle engagements took place. First, there was considerable epistemic transference *between* the two systems of thought. Second, global norms of order and sanction were problematized at the local level, at the same time as local forms of value and merit were rearticulated on much wider scales. Such developments remind us that struggles over science are geographical struggles also, be it in the way that forms of sanction are applied in and through space or be it in the way that geographical forms of sanction are themselves taken up as part of a broader vocabulary of suspicion or trust. We have seen how the US embargo of Cuba works in both of these senses.

The above tells us not just about the history but about the geography of reason and resistance within the biotechnology mode of production more broadly. In elaborating this geography, I have tried to argue that the articulation of particular scientific forms over space is in certain important ways constitutive of those forms themselves. Cuba's biotechnology project emerged, for example, at a crucial juncture both for the global economy and for science itself. It emerged at a time when the biological sciences, through biotechnology, were becoming heavily capitalized, and when it seemed that in biotechnology there was no alternative, as indeed another British prime minister, Margaret Thatcher, had famously asserted about neoliberal capitalism itself. But Cuba's biotechnology project—along with not dissimilar experiences in countries such as India and Brazil—*was* an alternative, and one that has pointed up some of the contradictions inherent in Western science more broadly. Not least, as one British scientist I spoke to said, it has shown that "what is supposed to be 'do-able' [under pressure of economic and political decisions] becomes quite impossible."[2] As this book has tried to show, the alternative that Cuba offered was predicated on the particular ways that Cuban science was (and was not) connected to global science.

## Political Economy and Epistemology

It is possible to develop some of these geographical insights a little further. The events in this book problematize the relationship between science, space, and political-economic orders. They also take place in the context of prior discussions of ideology. For Marx, ideology exists wherever attention is diverted from its true subject. Science, by contrast, was that which he saw could tear through the veil of illusion that is ideology's only substance. For Marx, science and ideology were thus counterposed. Much writing about Cuban biotechnology fails to distinguish between the two, however. Supporters of Cuban science tend to assume that socialist ideology somehow promotes better science, while its critics assume that Cuban science is nothing but socialist ideology—perhaps even mere idealism. Both assumptions are false, of course, but we need also to turn to philosopher of science Georges Canguilhem to understand why.

In his reading of Marx, Canguilhem agreed that Marx was right to say that even pure science, though it might help to puncture ideology, is not immune from commercial and industrial demands. But it needs to be clarified just how. Canguilhem goes on to suggest that we ought to reject the notion of an *ideological science* (which he sees would always be a false science, and therefore have no history anyway—it would never have anything to prove), while still allowing for what he calls *scientific ideology*. Such scientific ideologies refer to what is believed to be true at any given point *within* a scientific discourse. Scientific ideologies are therefore entirely necessary to science. Gregor Mendel's eventual discoveries about genetics, for example, were possible only because he had initially hoped for and sought to establish a very different set of rules. "In scientific ideology," Canguilhem said, "there is a desire to be science, in imitation of some already constituted model of what science is [i.e., a scientific discourse]." A scientific ideology comes to an end when it is taken over by a discipline that operationally demonstrates the validity of its claim to scientific status. Canguilhem's point is simply that what a particular scientific practice demonstrates, what it defines, is not necessarily what its prior scientific ideology suggested looking for.

What Canguilhem suggests in relation to Marx provides some considerable clarity about the Cuban case. The first of the two sets of claims above, for example (that Cuban science is a specifically *socialist* science and is therefore better for it), and which I examined in the earlier chapters, is clearly an attempt to create an "ideological science," as Canguilhem refers to it, and so must be dismissed in terms of its bearing on epistemological practice. But so too must the second set of claims—that Cuban science is nothing but ideological posture masquerading as real difference in scientific thought and practice—because, as we have seen in later chapters, there were real differences in the Cubans' scientific work. The answer to where the difference (and value) in the Cubans' work comes from lies between these claims. It was the very contradictions between *competing* attempts to create an ideological science around Cuban biotechnology work (articulated primarily in terms of nationalist and socialist rubrics first of all, but later in dialogue also with capitalism) that provided the conditions of existence for a particular approach to science (for particular "scientific ideologies") to be realized, sometimes successfully becoming displaced by demonstrable scientific fact, sometimes not. So it is the attempts to produce ideological science that Marx focuses on, and the conditions for scientific ideology they bequeath to science that Canguilhem examines, that together help to explain the emergence of a specific Cuban biotechnology. Science does not therefore *overlay* scientific ideology (as Canguilhem ultimately claims); we should in fact say that it *displaces* it, and the attempts to produce ideological science, along with political-economic conditions of work such ideologies dialogue with, help determine *how* scientific ideology comes to be displaced by science. By examining the real nature of Cuban scientific work, we come to consider also, therefore, the process of making space for scientific thought.

In order to work in the world, Cuban discoveries had to achieve the status of scientific fact, and to do that they had to be "made to work" or "demonstrated" outside of the milieu in which they were created. As Canguilhem goes on to say, the production of new knowledge is thereby dependent as much upon the "nonscience" of its materialization as it is on the "science" of its discovery. But we cannot consider the two apart. For this reason, the history of the discovery of a solution to meningitis B, for example, cannot stop with its scientific elucidation. As a scientific ideology, the theory of how meningitis B could be vaccinated against had also to unfold in the world of ideologies of science, where any deviation from the norm—in terms of the forms of warrant used to verify the theory (such as the clinical trials undertaken in Cuba in the 1980s), say—was liable to be seen as veering toward "epistemic error and moral danger."[3] It is not by chance that epistemic error and moral danger have been, as we saw in chapter 7, two of the most common critiques leveled at the Cuban production of pharmaceuticals. Some science *is* ill-disciplined of course.[4] But so too are many of the determinations that go into deciding *which* science is ill-disciplined: determinations that can often be made on an *interpolated* as much as an *interpretative* basis. In the Cuban case we have seen, moreover, how such determinations were sometimes assumed on the basis of prior architectures of distrust: of socialism, of poverty, of Castro, of the "value" of Cuban drugs.

## Experimental Geographies

Such interpolations overlook what it is that Cuban research most consistently achieves: not the formation of a new concept, nor the manufacture of a new entity, so much as the demarcation of a novel structure; namely, a reconfiguration of the relations between things. In the process the Cubans have condensed a globally disparate set of approaches to a biopharmaceutical substance into a deliverable medicine (the application of interferon), they have made a governmental rationality an element of methodology (the scale of clinical trials required to "produce" the meningitis B vaccine) and they have turned an obstacle into a tool (the use of epidermal growth factor itself in anticancer therapies). They have not in every case made a scientific breakthrough. Quite often, as with their approach to intellectual property, it has simply been that they have had no choice but to work outside the norms of Western scientific orthodoxy. But at the same time they have also contributed an analytic work that sets the relations between people (competition, cooperation, obligation) alongside the relations between things (objects, problems, theories) in different ways than scientists working elsewhere.

Michel Foucault considers the elucidation of precisely this sort of epistemic geography. "That man lives in a conceptually architectured environment does not prove that he has been diverted from life by some oversight or that a historical drama has separated him from it; but only that he lives in a certain way, that he has a relationship with his environment such that he does not have a fixed point of view of it, that he can move on an undefined territory, that he must move about to receive information, that he must move things in relation to one another in order to make them useful."[5] Writing as he was in the introduction to Canguilhem's book *The Normal and the Pathological*, Foucault was thinking most specifically of that form of knowing that derives from life, or vitality, itself. In a related vein, I have sought to consider here how particular modes of life—of not "being" socialist, but working in and through socialism, of not "becoming" capitalist, but of relating to capitalism—are bound up in scientific endeavor.

It is here that the value of a geographical perspective becomes clear. "Science has to be credible and embodied if it is to be produced and sustained, and . . . it has to be spatially located," state the editors of a history of science volume entitled *Making Space for Science*.[6] More than simply being located, as I have been arguing, it is in fact the *connections* that a spatialized epistemology of science must concern itself with. It is not just places, contexts, and sites that we must interrogate but the social relations between them—and even more so the articulation of such things as reason and resistance as part of those social relations. As we have seen, these myriad connections work in and through different politics of scale; they run down some cultural gullies and not others; they take on different forms in different places. The "Cuban biotechnological"—examined here mainly through the prism of the Science Pole—may be a highly sanctioned space, a space of constraint. But it is also for the same reason one of freedom: there one explores most fully the freedom not to be free, the freedom not to be bound by the rules of the norm, be those norms set by a socialist state or by international regulatory organizations. The sort of scientific work that emerged in this space of possibility in Cuba was not itself intended to be innovative, but it became so.

But if such a spatialized epistemology as we have engaged with offers us a way of accounting for the Cuban experience, so too does that experience shed light on the geographical nature of knowledge itself. It tells us, for example, about the articulation of knowledge over space. The Cuban laboratories had their own internal division of labor and spatial organization of course, but this local scientific work was given shape in dialogue with events taking place elsewhere in political, scientific, economic, and social registers. The labs were not simply sited *within* these broader geographies. There was a scaled geography at work. International norms were partially articulated in a national setting that was further reworked through the regional culture of the Science Pole and then again at the local level of the laboratory. All of this allows for the variations, mutations, and reformulations of scientific "standards," "norms," and "values" that were such an important part of the formulation of a novel epistemic milieu.

It tells us also about the articulation of space within knowledge. If truth is simply "what science speaks"—as historians and philosophers of science tell us—then it surely warrants asking "which science?" and "from where does it speak when it speaks to us?" Temporal assumptions lie at the heart of many philosophies of science: "Every historian of science is necessarily a historiographer of truth," Gaston Bachelard says.[7] But since scientific knowledge is also constituted geographically, attempts to mobilize scientific knowledge founder outside of a simultaneous mobilization of the world. Thus there was a particular geography that was mobilized first in Casa 149: not just Finnish purifications and secondhand American equipment but a form of international solidarity that gelled them together. And while it might have worked just as well in the laboratory if the Cubans were following an American method for purification, that it was Finnish helped mobilize their results in the particular (then Soviet) world that was required.

Finally, it tells us something of the way that space is remade in the process of epistemic exchanges. Whether it be new institutions called forth to support or corral emergent scientific efforts, or the need for new regulatory systems in response to unexpected scientific advances, new spaces are brought forth and in turn make possible new forms of social activity. When one innovates, one also contributes spatial innovations. New scientific facts (such as the "role" of EGF in the reproduction of cancer cells) produce new spaces within which those facts are given meaning, operate, are circulated, and so on. The production of new knowledge requires, therefore, not only conceptual developments but also enabling technologies and, most of all, the enabling of the environment in which they will unfold. Hence the Cubans' work on cancer revealed the value of a different epistemological structuring of a particular scientific problem in Cuba as compared to elsewhere; it was the product of Cuban scientists producing a particular geography of work as they sought to make innovations in scientific practice and thought. Similarly, as the Cubans worked on *neisseria* bacteria in relation to meningitis B, a new space for thought opened up for subsequent work on cancer that might use those bacteria as an adjuvant.

It would appear then that if science is overdetermined by capital, then so too is that relationship overdetermined by geography (and indeed, Louis Althusser's very notion of overdetermination was intended to denote a contextual relationship, not a linear or causal one).[8] Such geographies are not simply given, however; they are the product of work: they must be imagined, they must be literally "put into place," and they must be struggled over, usually all at the same time. Work goes into making some scientific forms more persistent than others, but at the same time—as the Cuban case shows—the politics of distance can sometimes make a privilege of peripherality.

## The Cuban Cure

If the Cuban experience is illuminative of certain of these deep-seated problems concerning the nature of biotechnology science in a modern, globalizing world, it is important, too, in a more tangible sense. Cuba's policy of putting biotechnology to work within a public health framework, focused on preventive medicine and often tied to a mission orientation, might seem rather incongruous, but it has worked for the reasons I have tried to describe in this book. Despite the difficulties of the Special Period, the country has maintained a relatively high level of public health and has continued to export some of its most basic life-enabling technologies to other countries in need.

The conditioning of life-science research under social and moral obligation of public interest has resulted in a very different life-science industry on that island. It is not one that all those scientists involved have agreed on, nor is it one free of certain restrictions or political enforcements which run contrary to its ethic and which have at times threatened to undermine it. Cuban biotechnology does not offer a model that other countries might unproblematically emulate, therefore.[9] But as with the not entirely dissimilar experiences in Brazil and in India, among other places, there *is* a good deal to learn from their experience.

None of these countries have found the answer to the problem of an overcapitalized biotechnology industry. But the Cuban case in particular is illuminating for the way that, in certain significant respects, it has challenged that system. Vaccines—pharmaceutical products that protect lives rather than merely enhance them—represent only 29 percent of the worldwide biotechnology pipeline; in Cuba—stripped of funding under an embargo—they represent 50 percent of the biotechnology pipeline.[10] This, then, is the cure that Cuba might be said to offer. No magic bullet to be sure, but an attempt, at least, to produce a more social science.

# NOTES

### INTRODUCTION

1. *Bohemia* (July 1986): 4. All translations are by the author except when otherwise noted.
2. Atchuk Tcheknovorian, cited in *Granma*, July 2, 1986.
3. Rodolfo Quintero Ramírez, cited in *Granma*, April 21, 1989.
4. Fidel Castro, cited in David Lipschultz and Peter Rojas, "The Next Biotech Corridor: Cuba?," *Red Herring* magazine, April 12, 2001.
5. Fernando Ortíz, *Cuban Counterpoint: Tobacco and Sugar*, trans. Harriet de Onís (1940; Durham, NC: Duke University Press, 1995).
6. Francis Fukuyama, *Our Posthuman Future: Consequences of the Biotechnology Revolution* (London: Profile Books, 2002).
7. Ernst and Young Global Biotechnology Report, 2006.
8. See Martin Kenney, *Biotechnology: The University Industry Complex* (New Haven, CT: Yale University Press, 1986), and Steven Wright, "Recombinant DNA Technology and Its Social Transformation, 1972–1982," *Osiris* 2, 2nd ser. (1986): 303–60. The program for the BIO 2006 Annual International Conference can be seen at http://www.softconference.com/260409.
9. Jikun Huang et al., *Science* 295, no. 5555 (Jan. 25, 2002): 674–76.
10. See H. H. Gerth and C. Wright Mills, *From Max Weber: Essays in Sociology* (1948; London: Routledge and Kegan Paul, 1998), 129.
11. Fidel Castro, cited in *Granma*, July 2, 1986.
12. Thomas Kuhn, *The Structure of Scientific Revolutions* (1962; Chicago: University of Chicago Press, 1996).
13. I am aware of the pitfalls of spatial metaphors. Alice Jenkins ("Spatial Rhetoric in the Self-Presentation of Nineteenth-Century Scientists: Faraday and Tyndall," in *Making Space for Science: Territorial Themes in the Shaping of Knowledge*, ed. Jon Agar, Crosbie Smith, and Gerald Schmidt [London: Palgrave Macmillan, 1998], 181) sets this out quite clearly: "Spatial metaphors produce a dual movement in texts. They act subtly to draw language away from the objective or the external into fiction; but they are often used in an endeavour to concretize the abstract." I have tried to be careful.
14. Nigel Thrift et al., "Editorial: The Geography of Truth," *Environment and Planning D: Society and Space* 13, no. 4 (1995): 3.

15. There are numerous examples one could cite here, but a good starting point is Agar, Smith, and Schmidt, eds., *Making Space for Science*.
16. David Livingstone, *Putting Science in Its Place: Geographies of Scientific Knowledge* (Chicago: University of Chicago Press, 2003). For certain implicit geographical themes, see the special issue of *Social Studies of Science* on postcolonial technoscience, esp. W. Anderson, "Introduction: Post-Colonial Technoscience," *Social Studies of Science* 32, nos. 5–6 (2002): 643–58.

CHAPTER ONE

1. Jean Lindenmann et al., "Studies on the Production, Mode of Action and Properties of Interferon," *British Journal of Experimental Pathology* 38 (1957): 551–62.
2. See Jean Lindenmann, "How To Avoid Making a Fortune in Medicine," *Nature* 394, no. 6696 (1998): 844–45.
3. The idea of a "magic bullet" was first proposed by Paul Erlich who, at the beginning of the twentieth century, figured that if a compound could be made that selectively targeted a disease-causing organism, then a toxin for that organism could be delivered along with the agent of selectivity. Only two other antivirals were licensed for use at the time. See D. Kinchington, "Recent Advances in Antiviral Therapy," *Journal of Clinical Pathology* 52 (1999): 89–94.
4. Donald Holmquest, letter to R. Lee Clark, October 16, 1980, R. Lee Clark archive, Austin, Texas.
5. Randolph Lee Clark papers, R. Lee Clark archive, Austin, Texas.
6. Clark's visit coincided with the peak of media hype about interferon. See, e.g., Kari Cantell, *The Story of Interferon: The Ups and Downs in the Life of a Scientist* (London: World Scientific Publishing Company, 1998), 146.
7. The "interferon group," as they would become known, consisted of doctors Limonta and Ramírez, virologists Silvio Barcelona and Ángel Aguilera, and biochemists Eduardo Pentón and Pedro López-Saura. In Cuban Spanish, *cuadro* denotes a dedicated professional, often working within a specific unit.
8. Pedro López-Saura, interview, June 6, 2002.
9. Lindenmann, "How To Avoid Making a Fortune in Medicine."
10. Those side effects were potentially severe. In France in 1982, two people died after having received a dose of impure natural leukocyte interferon.
11. D. M. Dorta, *Impacto del IPK en el control de Aedes aegypti en Santiago de Cuba durante la epidemia de dengue y en ciudad de la Habana* (Havana: Instituto de Medicina Tropical "Pedro Kourí," 1987). See also Gustavo Kourí, G. Guzmán, et al., "Hemorrhagic Dengue in Cuba: History of an Epidemic," *Pan American Health Organization Bulletin* 20 (1986): 24–30.
12. Dorta, "*Impacto del IPK.*" Dengue is an infectious disease produced by a genomic virus and for which the *Aedes aegypti* mosquito is the principal transmission vector.
13. See, e.g., Kourí et al., "Hemorrhagic Dengue in Cuba."
14. See the various articles penned by the interferon group in scientific journals: Manuel Limonta, V. Ramírez, et al., "Uso del interferón leucocitario durante una epidemia de dengue hemorrágico (virus tipo II) en Cuba," *Interferón y Biotecnología* 1, no. 3 (1984): 15–22; Gustavo Guzmán, G. Kouri, et al., "Inhibición de la multiplicación del virus dengue en presencia de interferón," *Interferón y Biotecnología* 4, no. 2 (1987): 108–14; Victoria Ramírez, A. Gonzáles Griego, et al., "Uso del interferón-a leucocitario por vía intraperitoneal en humanos: Aspectos farmacocinéticos," *Inter-*

*ferón y Biotecnología* 1, no. 1 (1984): 31–40; A. Sotto, E. Selman, et al., "L'interferon leucocitaire dans les hépatitis virales subaigues," *Med.Chir.Dig.* 15 (1986): 103–5.

15. Michael Fransman, telephone interview, January 16, 2001; emphasis added. Fransman is a former member of the UN University INTECH program.
16. The initial call for a moratorium on research was M. F. Singer and D. Soll, "Guidelines for Hybrid DNA Molecules," *Science* 181 (1973): 1114.
17. Paul Rabinow, *Making PCR: A Story of Biotechnology* (Chicago: University of Chicago Press, 1996), 17; emphasis added. This, and his *French DNA* (Chicago: University of Chicago Press, 1999), are genre-defining ethnographies which helped me develop my own approach to Cuban biotech.
18. Carlos Borroto, "Biotecnología agropecuario," in *Cuba: Amanecer del tercer milenio,* ed. F. C. Díaz-Balart (Havana: Editorial Científico Técnica, 2002), 26.
19. Hacking, Ian, "The Self-Vindication of the Laboratory Sciences," in *Science as Practice and Culture,* ed. A. Pickering (Chicago: University of Chicago Press, 1992). See also Sharon Traweek, *Beantimes and Lifetimes: The World of High Energy physicists* (Cambridge, MA: Harvard University Press, 1992).
20. Michel Foucault, in *The Foucault Reader,* ed. Paul Rabinow (Harmondsworth: Penguin, 1984), 85. This is in contrast, e.g., to Cantell's comment that, "clearly Castro himself was behind it [the biotechnology project]" (Cantell, *Story of Interferon,* 145). While Cantell is right, he also misses the point.
21. Philip Barker, *Michel Foucault: An Introduction* (Edinburgh: Edinburgh University Press, 1998), 21.

CHAPTER TWO

1. A similar and perhaps more readily available impression can be gained from Alberto Elena and Javier Ordóñez, "Science, Technology, and the Spanish Colonial Experience," in *Nature and Empire: Science and the Colonial Enterprise,* ed. Roy MacLeod, *Osiris* 15 (Chicago: University of Chicago Press, 2001), 70–84. There is a rich body of work in the history of science dealing with science and medicine, colonialism and space. For example, David Arnold, *Colonizing the Body: State Medicine and Epidemic Disease in Nineteenth-Century India* (Berkeley: University of California Press, 1993), and Peter Redfield, *Space in the Tropics: From Convicts to Rockets in French Guiana* (Berkeley: University of California Press, 2000).
2. Loren Graham, ed., *Science and the Soviet Social Order* (Cambridge, MA: Harvard University Press, 1990), 1. See also Graham's *Science, Philosophy, and Human Behavior in the Soviet Union* (New York: Columbia University Press, 1987).
3. An informative overview is provided by one of Cuba's eminent historians of science, José López Sánchez, *Ciencia y medicina: Historia de las ciencias* (Havana: Editorial Científico Técnica, n.d.).
4. Finlay studied the *Stegomyia fasciata* mosquito, referring to it as the Culex mosquito.
5. "Carlos Finlay y la fiebre amarilla," Juan A. del Regato Collection, box 2, Cuban Heritage Collection (henceforth CHC), Miami. See esp. François Delaporte, *The History of Yellow Fever: An Essay on the Birth of Tropical Medicine* (Princeton, NJ: Princeton University Press, 1991). But see also Tirso Sáenz and E. G. Capote, *Ciencia y tecnología en Cuba: Antecedentes y desarrollo* (Havana: Editorial de Ciencias Sociales, 1989); Fernando Ortíz, *La hija cubana del iluminismo* (Havana: Editorial Academica 1993 [1943]); and José López Sánchez, "Breve historia de la ciencia en Cuba," *Revista de la Biblioteca Nacional José Martí* 7, no. 1 (1980): 21–49.

6. On the creation of the Cuban Academy of Sciences in relation to nationalist struggles against Spain, see Pedro M. Pruna, "National Science in a Colonial Context," *Isis* 85 (Sept. 1994): 412–26.
7. R. J. Paquette, *Sugar Is Made with Blood: The Conspiracy of La Escalera and the Conflict between Empires over Slavery in Cuba* (Middleton, CT: Wesleyan University Press, 1988). See also Louis Pérez Jr., *Cuba under the Platt Amendment, 1902–1934* (Pittsburgh, PA: University of Pittsburgh Press, 1986), and Jorge Ibarra, *Prologue to Revolution: Cuba 1898–1958* (London: Lynne Rienner, 1998).
8. Oscar Pino-Santos, *El asalto a Cuba por la oligarquía financiera yanqui* (Havana: Casa de las Américas, 1973).
9. International Bank for Reconstruction and Development (IBRD), "Report on Cuba: Findings and Recommendations of an Economic and Technical Mission Organized by the International Bank for Reconstruction and Development in Collaboration with the Government of Cuba in 1950" (Baltimore, MD: The John Hopkins Press, 1951), CHC. See also Luis Herrera, "Ingeniería genética," in *Cuba: Amanacer al tercer milenio*, ed. F. C. Díaz-Balart (Havana: Editorial Científico Técnica, 2002). Herrera is the current director of Cuba's Center for Genetic Engineering and Biotechnology.
10. Kitty Abraham, "Postcolonial Science, Big Science and Landscape," in *Doing Science + Culture*, ed. Sharon Traweek and Roddey Reid (London: Routledge, 2000).
11. Fernando Coronil, *The Magical State: Nature, Money, and Modernity in Venezuela* (Chicago: University of Chicago Press, 1997), 6. For a not dissimilar argument focusing on science, and Spanish colonial science in particular, see Jorge Cañizares-Esguerra, "Iberian Colonial Science," *Isis* 96 (2005): 64–70.
12. Which is how it appears in Benedict Anderson's original account: Benedict Anderson, *Imagined Communities: Reflections on the Origin and Spread of Nationalism* (London: Verso, 1983).
13. A. Korbonski and F. Fukuyama, eds., *The Soviet Union and the Third World: The Last Three Decades* (Ithaca, NY: Cornell University Press, 1987); see also Ankie Hoogvelt, "Prospects in the Periphery for National Accumulation and Liberation in the Wake of the Cold War and Debt Crisis," in *Regimes in Crisis: The Post-Soviet Era and the Implications for Development*, ed. B. Gills and S. Qadir (London: Zed Books, 1995), and Ankie Hoogvelt, *Globalization and the Post-Colonial World: The New Political Economy of Development* (Baltimore, MD: The Johns Hopkins University Press, 1997).
14. See Paul Josephson, "Soviet Scientists and the State: Politics, Ideology, and Fundamental Research from Stalin to Gorbachev," *Social Research* 59, no. 3 (1992): 589–614; see also Bruce Parrott, *Politics and Technology in the Soviet Union* (Cambridge, MA: MIT Press, 1985).
15. Something perhaps most clearly articulated long ago by Joseph Dietzgen: "Modern socialism . . . is scientific. Just as scientists arrive at their generalizations not by mere speculation, but by observing the phenomena of the material world, so are the socialistic and communist theories not idle schemes, but generalizations drawn from economic facts." Joseph Dietzgen, *Philosophical Essays* (1873; Chicago: C. H. Kerr & Company, 1917), n.p.
16. E. Guevara, *Nuestra industria económica* 13 (Havana: June 1965), Biblioteca Nacional José Martí (henceforth BNJM).
17. F. Castro, *Granma*, April 20, 1968, BNJM. He did not mention that many electrical goods sourced from the Soviet Union were incompatible with the US-built electricity grid anyway: DRE: "*Historia de un desastre económico*" (Caracas, Venezuela: 1963), 52, BNJM.

18. Antonio Nuñez-Jiménez, ACC, Serie Isla de Pinos, no. 1, Havana. Nuñez-Jiménez was the first president of the reestablished Cuban Academy of Sciences and Cuba's preeminent geographer.
19. Philip Corrigan, Harvie Ramsey, and Derek Sayer, *For Mao: Essays in Historical Materialism* (London: Macmillan, 1979). These ideas have been more explicitly outlined in Antonio Nuñez-Jiménez, *Hacía una cultura de la naturaleza* (Toward a culture of nature; 1998). Such ideas strike a resonance both with contemporary notions of social capital, on the one (material) hand, and the social theory of hybridity, on the other (conceptual) hand.
20. See Fidel Castro, *Discurso pronunciado en la inauguración de la programa de UNESCO para 1969–70* (Havana: Ediciones COR, 1968); and Fidel Castro, *Discurso pronunciado en la clausura del primer congreso del Instituto de Ciencia Animal, el 13 de mayo de 1969* (Havana: Ediciones COR, 1969), no. 8. The strength of Cuba's national scientific tradition, recently rearticulated under the revolution, provided a certain distance from the worst excesses of Soviet science. While the structural organization of Cuban science in the 1960s may have taken the Soviet experience into account, therefore, a space for Cuban independence was retained. In 1976, for example, at the height of Cuban-Soviet cooperation, Cuban scientific institutions were also affiliated with forty-eight other international scientific organizations, and a further thirty-three applications were being processed (MINSAP, *Informe Annual*, Havana, 1976).
21. Sáenz and Capote, *Ciencia y Tecnología en Cuba*, 74.
22. P. Gummett and J. Reppy, *The Relations between Defense and Civil Technologies* (Dordrecht: Klewer, 1987).
23. Cuban Academy of Sciences, *Informe Annual*, 1988.
24. Fidel Castro, *Ciencia, tecnología y sociedad, 1959–1989* (Havana: Editora Política, 1990), 23.
25. John Dewey, *Democracy and Education: An Introduction to the Philosophy of Education* (New York: Free Press, 1997), 239. Dewey also wrote that science was "the method of emancipating us from enslavement to customary ends, the ends established in the past" (John Dewey, *Essays in Experimental Logic* (1917; New York: Dover Books, 1953), 441. Paul Rabinow uses this same quote in a similar way in his *Making PCR*.
26. Castro later observed that, "the fact of our having more than 100 scientific institutions and thousands of scientists already working in these centres is something which we had not even conceived of in the first years of the Revolution." Transcript of interview with Tad Szulc (CHC, UM, box 3, Tad Szulc collection, 88).
27. JUCEPLAN, the Central Planning Board, is the Cuban equivalent of the Soviet GOSPLAN.
28. Hugh Thomas, *Cuba: The Pursuit of Freedom* (London: Eyre and Spottiswoode, 1971), 980.
29. In the 1971–75 five-year plan (*quinquenio*), 40 percent of state investment was in agriculture and less than 15 percent in industry; by the 1976–80 five-year plan, investment in industry had risen to 35 percent while investment in agriculture had fallen to 18 percent. Cuban Academy of Sciences, *Estado Actual* (Ministry for Science, Technology, and the Environment, October 1981), CITMA.
30. Cuban Academy of Sciences, *Estado Actual* (1985): 11, CITMA. One outcome of this was the introduction in 1973 of the Economic Management and Planning System (*Sistema de Dirección y Planificación*—SDPE), modeled on the 1965 Soviet economic reforms.

31. *Producto Social Global* (PSG). PSG is the Cuban calculation for Gross Domestic Product (GDP).
32. Cuban Academy of Sciences, *Estado Actual* (1985): 11, CITMA.
33. Universidad de La Habana, *Aspectos Organizativos*, 1975, BNJM.
34. There is no mention of "science and technology" in development planning documents until this point.
35. "... a significant achievement is the realization of a scientific and technical system capable of rapidly responding to urgent social problems such as attack from diseases introduced against our country," Cuban Academy of Sciences, *Estado Actual* (1985): 52, CITMA. This is a reference to what official history in Cuba records as a number of instances of biological warfare conducted against the country by the United States.
36. Ross Danielson, *Cuban Medicine* (New Brunswick, NJ: Transaction Books, 1979); Sergio Díaz-Briquets, *The Health Revolution in Cuba* (Austin: University of Texas Press, 1983); see also MINSAP, *Cuba: La salud en la revolución* (Havana: Editorial Orbe, 1975) and MINSAP, *Análisis del sector salud en Cuba* (Havana: Colaboración de OPS/OMS, 1996).
37. DRE, *Cuba: Historia de un desastre económico* 1963, 17, BNJM. See also Sáenz and Capote, *Ciencia y tecnología en Cuba*, 212, for the Cuban version of events.
38. José Ramon Machado Ventura, transcript of interview with Tad Szulc. Ventura was the third director of MINSAP, and under his directorship the health system took on this feature which would characterize it throughout the revolutionary period (CHC, UM, box 3, Tad Szulc collection). See also Cuba, *Anuario estadístico de salud*, 2001, 159, which gives an annual breakdown of the number of graduating medics.
39. Fidel Castro, transcript of interview with Tad Szulc (CHC, UM, box 3, Tad Szulc collection).
40. See figures, "Decline in Total Infectious Diseases and Growth in Malignant Tumours, 1970–2000," in *Anuario estadístico de salud*, 2000, 54. Malaria, diphtheria, and polio were all practically eradicated during the first half of the 1960s: *La salud publica en Cuba*, 1999, 32.
41. Díaz-Briquets, *Health Revolution in Cuba*, 104.
42. Anonymous public health specialist, interview, Havana, May 16, 2002.
43. The lack of trained personnel was, according to Daniel Goldstein, one of the main brakes on biotechnology's development in Latin America (cited in R. Gonzales and P. Durán, "Comentarios sobre eventos," *Interferon y Biotecnología* 6, no. 2 [1989]: 205–17). See also Daniel Goldstein, "Third World Biotechnology, Latin American Development, and the Foreign Debt Problem," in *Biotechnology in Latin America: Politics, Impacts, and Risks*, ed. N. P. Peritore and A. K. Galve-Peritore (Wilmington, DE: American Silhouettes, 1995).
44. Danielson, *Cuban Biomedicine*.
45. In 1973, for example, 484 million pesos were spent on education, and 400 million pesos were spent on health. See "El desarrollo económico cubano: experiencias y perspectivas," JUCEPLAN, August 1976, BNJM. Graduation figures from CEE 1981, 17, CITMA.
46. José de la Fuente, former vice director of the CIGB, telephone interview, January 23, 2002: "[Biotech] is not something that can happen without certain things in place. Firstly Cuba always had good medicine, not just after the revolution but before, in depth if not in breadth. By the 1980s there were several research institutions which

provided a critical mass of scientists." See also Ilizastigui Dupuy, *Salud, medicina y educación medica* (Havana: Editorial Ciencias Médicas, 1985).
47. This quote comes from a *Granma* editorial, July 2, 1994.
48. Ibid.
49. Damian Fernandez, *Cuba and the Politics of Passion* (Austin: University of Texas Press, 2000).
50. Fidel Castro, Foreign Broadcast Information Service (henceforth FBIS), 1984-12-29; Castro's speeches are documented at the FBIS Castro Speech Database, available at http://lanic.utexas.edu/la/cb/cuba/castro.html.
51. E. R. Borroto Cruz and T. R. Medrano, *Descriptive and Comparative Study of the Population's Perception and Satisfaction with the Doctor-Patient Relationship in Havana City* (Havana: Higher Institute of Medical Studies, University of Havana, 1990). See also B. Roca, *Los fundamentos del socialismo en Cuba* (Havana: Ediciones Populares, 1990). Other values, such as "health internationalism" (as mentioned, e.g., in Julie Feinsilver, *Healing the Masses: Cuban Health Politics at Home and Abroad* (London: University of California Press, 1993), are an expression of this. For example, by 1985, Cuba had more medics working abroad than the WHO itself (PAHO, 2000).
52. Jon Anderson offers perhaps the clearest discussion of this in his biography of Che Guevara: Jon Lee Anderson, *Che Guevara: A Revolutionary Life* (London: Bantam, 1997), 636.
53. H. H. Gerth and C. Wright Mills, *From Max Weber: Essays in Sociology* (1948; London: Routledge and Kegan Paul, 1998).
54. CITMA, *Estudio prospectivo* (1982), CITMA.
55. Peter Schwab, *Cuba: Confronting the US Embargo* (Basingstoke: Macmillan, 1999), 72.
56. Georges Canguilhem, *Ideology and Rationality in the History of the Life Sciences* (Cambridge, MA: MIT Press, 1998), 5.
57. Fidel Castro, transcript of interview with Tad Szulc (CHC, UM, box 3, Tad Szulc collection, 4b).

CHAPTER THREE
1. Pedro López-Saura, vice director for regulatory affairs, CIGB, interview, Havana, June 6, 2002.
2. "UNIDO World Bio-Center Meeting Puts Off Decisions," *Biotechnology Newswatch* 5, no. 8 (April 15, 1985): 7.
3. Harry Collins and Robert Evans, "The 'Periodic Table' of Expertises," 2004, www.cfa.ac.uk/socsi/expert. On page 8 they say: "The more distant is one from the locus of creation of knowledge in social space and time the more certain will the knowledge appear to be."
4. Julie Feinsilver, *Healing the Masses: Cuban Health Politics at Home and Abroad* (Los Angeles: University of California Press, 1993), 151.
5. Martin Fransman, former member of the UN University INTECH program, telephone interview, January 16, 2001. Fransman visited Cuba in the 1980s in connection with the UNIDO center.
6. Feinsilver, *Healing the Masses*, 136. In a 1986 report in *Biotechnology Newswatch*, an international industry review publication, Dr. Silvio Barcelona, a director of the CIB and member of the original team, stated at the time that "for MediCuba to advertise the recombinant interferon is an error," and likewise Dr. Alejandro Silva, head

of the microbiology laboratory at the CIB, confirmed that "it's only been tested in animals so far. We need more results before commercializing this interferon" (*Biotechnology Newswatch* 6, no. 6 [March 17, 1986]).

7. Fidel Castro, Foreign Broadcast Information Service (henceforth FBIS): 1985-09-29, available at http://lanic.utexas.edu/la/cb/cuba/castro.html.
8. Antoni Kapcia, *Cuba: Island of Dreams* (Oxford: Berg, 2000), 205.
9. Julio García-Luis, *Cuban Revolution Reader: A Documented History* (Melbourne: Ocean Press, 2001), 242.
10. Kapcia, *Cuba: Island of Dreams*, 207.
11. Ibid.; emphasis added.
12. This is a point I take from Stuart Corbridge, *Marxisms, Modernities, and Moralities: Development Praxis and the Needs and Rights of Distant Strangers* (Department of Geography, Cambridge University: offprints collection, 1993). See also Manuel Castells, *The Rise of the Network Society*, vol. 1 of *The Information Age: Economy, Society and Culture* (London: Basil Blackwell, 1996).
13. Cuban Academy of Sciences, *El Estado Actual*, 1981, 55.
14. ExpoCuba, 1989: 1. In his series of interviews with Cuban biotech leaders, the Cuban scientist Ernesto Bravo makes the same point (*Development within Underdevelopment: New Trends in Cuban Medicine* [Havana: Editorial José Martí/Elfos Scientiae, 1998]).
15. I am grateful to one of the manuscript reviewers for clarification here.
16. Economist Intelligence Unit, *Country Report on Cuba*, 2001.
17. In *Palabras a los intelectuales*—Words to the intellectuals—Castro famously asserted: "dentro de la revolución todo; contra la revolución, nada"—within the revolution, everything; against the revolution, nothing." Fidel Castro, *Palabras a los intelectuales* (1961), 20.
18. Cuban Academy of Sciences, *El Estado Actual*, 1981, 51.
19. Account given by of one of the builders, Jesús Lomba Mestre, quoted in *Granma*, July 2, 1986.
20. Manuel Limonta, "II seminario cubano sobre interferón y I seminario cubano sobre biotecnología" in "IFN Cuba," Memorias, part 1, February 20-22 (Havana: CIGB, 1986): 35.
21. These ideas are elaborated in Manuel Limonta, "Biotechnology and the Third World: Development Strategies in Cuba," in *Biomedical Science and the Third World: Under the Volcano*, ed. B. Bloom and A. Cerami (New York: New York Academy of Sciences, 1989), 325-34.
22. "Este centro no sólo va a ser vanguardía en Biotecnología, sino va a ser vanguardía entre los centros de investigación científica de nuestro país" (Fidel Castro, speech at the opening of the CIGB, cited in *Granma*, July 2, 1986). And on January 8, 1989, Castro described the CIGB as itself being "a vanguard technology which will determine the future development of Cuba"; Fidel Castro, *Discurso pronunciado en el acto central por el aniversario 30 de su entrada a La Habana*, *Granma*, January 11, 1989.
23. Such international exchanges were, according to José de la Fuente, "important components of the [center's] success during the early development of Cuba's biotechnology" (José de la Fuente, "Wine into Vinegar: The Fall of Cuba's Biotechnology," *Nature Biotechnology* 19 (2001): 905-7.
24. See Albert Sasson, "Biotechnologies—A Problem for Society," *Recherche* 14, no. 147 (1983): 1156-61; Albert Sasson, "Biotechnology—Challenges and Prospects," *Journal of Scientific and Industrial Research* 44, no. 4 (1985): 171-85; and Albert Sas-

son, "Cuban Biotechnology," *Biofutur* 52 (1986): 65–66. See also Albert Sasson, "Biotechnologies and the Use of Plant Genetic Resources for Industrial Purposes: Benefits and Constraints for Developing Countries," *Biotechnology and Development Review* 4 (1995): 1–14.
25. Agustín Lage, "Las biotecnologías y la nueva economía: crear y valorizar los bienes intangibles," *Biotecnología Aplicada* 17 (2000): 59.
26. For an analysis of the ways that landscapes are read as texts, see Jim Duncan and Trevor Barnes, *Writing Worlds: Discourse, Text, and Metaphor in the Representation of Landscape* (New York: Routledge, 1992).
27. Fidel Castro, FBIS: 1992-00-62, available at http://lanic.utexas.edu/la/cb/cuba/castro.html.
28. James Dearing, *Growing a Japanese Science City: Communication in Scientific Research* (London: Routledge, 1995).
29. See the analysis set out in Doreen Massey, Paul Quintas, and David Wield, *High-Tech Fantasies: Science Parks in Society, Science and Space* (Routledge: London, 1992).
30. F. R. Rodríguez, "El papel de la planifacación en la economía nacional," *Revista Interamericana de Planificación* 8 (1973): 102–15. See also S. C. Park, "Globalization and Local Innovation System: The Implementation of Government Policies to the Formation of Science Parks in Japan and South Korea," *Korea Observer* 31, no. 3 (2000): 407–48, and D. H. Shin, "An Alternative Approach to Developing Science Parks: A Case Study from Korea," *Papers in Regional Science* 80, no. 1 (2001): 103–12 for other examples of public-sector-led science park development, particularly in Japan and South Korea.
31. Dearing, *Growing a Japanese Science City*, 39.
32. Chandra Mukerji, *A Fragile Power: Scientists and the State* (Princeton, NJ: Princeton University Press, 1989).
33. Timothy Lenoir, *Instituting Science: The Cultural Production of Scientific Disciplines* (Stanford, CA: Stanford University Press, 1997), 75.
34. A useful discussion of Weber's treatment of Tolstoy may be found in David Owen, Tracy Strong, and Rodney Livingstone, eds., *Max Weber: The Vocation Lectures: Science as a Vocation, Politics as a Vocation* (Indianapolis, IN: Hackett Publishing Company, 2004).

CHAPTER FOUR

1. Traweek makes a similar point here: Sharon Traweek, "Border Crossings: Narrative Strategies in Science Studies and among Physicists in Tsukuba Science City, Japan," in *Science as Practice and Culture*, ed. Andrew Pickering (Chicago: University of Chicago Press, 1992), 429–66.
2. Robert K. Merton, *The Sociology of Science: Theoretical and Empirical Investigations* (Chicago: University of Chicago Press, 1973).
3. Sierra was one of the principal researchers who created the VA-MENGOC-BC.
4. "U.S. Finally Will Let SmithKline Market Cuban Meningitis Vaccine," *Wall Street Journal*, July 23, 1999.
5. Gustavo Sierra and Concepción Campa, then at the National Center for Meningitis Vaccine of MINSAP, *Interferon y Biotecnología* 6, no. 2 (1989).
6. In a paper published in the journal *NIPH Annals* in 1991, the Cuban researchers describe this process: "A randomized, double-blind controlled trial of the vaccine efficacy was performed during 1987–1989 with 106,000 10–14 years students from 197 boarding schools in seven provinces." The efficacy then obtained was 83 per-

cent. In the subsequent rollout of the actual vaccine, the efficacy ranged from 83 to 94 percent in different provinces. See G. V. G. Sierra, H. C. Campa, N. M. Varcacel, et al., "Vaccine against Group B *neisseria meningitides*: Protection Trial and Mass Vaccination in Cuba," *NIPH Annals*, 14, no. 2 (Dec. 1991): 195–207.

7. Ibid., 195.
8. For a full comparative analysis of Norwegian and Cuban vaccine research, see Jens Plahte, "Vaccine Innovation for Low-Revenue Markets: The Cuban Vaccine Industry in a National and Global Context," PhD dissertation, Center for Technology, Innovation, and Culture, University of Oslo, 2009. I am grateful to Jens for many illuminating conversations on Cuba.
9. "Insignia de la Ciencia Cubana," typescript, CITMA, personal copy. The employees at the Finlay Institute are certainly rewarded for such dedication: in 2002, Sierra told me, they received double the national average wage. Meals and childcare are provided, as at the CIGB, and twice-yearly bonuses are given in US dollars.
10. Cited from an interview with *Medicc Review*, Havana, 2004.
11. Fidel Castro, Foreign Broadcast Information Service (henceforth FBIS): 1992-02-45, available at http://lanic.utexas.edu/la/cb/cuba/castro.html.
12. AnnaLee Saxenian, *Regional Advantage: Culture and Competition in Silicon Valley and Route 128* (London: Harvard University Press, 1994), 30. These comments are not limited to the US context. The success of Japanese industry and business is also attributable in part to network organizational forms: see Jon Sigurdson and A. M. Anderson, *Science and Technology in Japan* (Harlow: Longman, 1991), and B. C. L. Christenson, "The Third China? Emerging Industrial Districts in Rural China," *International Journal of Urban and Regional Research* 21, no. 4 (1997). The Japanese corporation is more internally decentralized and more open to the surrounding economy than the traditional large American corporation.
13. Studies of which may be found in Edward Goodman, Julia Bamford, and Peter Saynor, eds., *Small Firms and Industrial Districts in Italy* (London: Routledge, 1990), A. Scott, *New Industrial Spaces: Flexible Production Organization and Regional Development in North America and Western Europe* (London: Pion, 1988), and M. Storpor and A. Scott., eds., *Pathways to Industrialization and Regional Development* (London: Routledge, 1992).
14. Saxenian, *Regional Advantage*, 4; though see Bennett Harrison, *Lean and Mean: Why Large Corporations Will Continue to Dominate the World Economy* (New York: Guildford Press, 1998), 23, for an insightful critique.
15. E. Dore, *Flexible Rigidities* (London: Athlone Press, 1986); C. Mukerji, *A Fragile Power: Scientists and the State* (Princeton, NJ: Princeton University Press, 1989).
16. Traweek, "Border Crossings."
17. Source, José Ramón Fernandez, vice president of the Council of Ministers, Ministry of Education of the Republic of Cuba, 1986, "*Discurso Inaugural*" at the IFN symposium. Labor figures are not broken down by job type in these figures but by the ministry to which individuals pertain. Other sources put the numbers as slightly higher: 40,000 workers of all sorts in Cuba slightly before this time.
18. This figure appeared in a number of interviews. It corroborates closely with the average age given by a US Center for Defense Information (CDI, 2003) survey of nine institutes of the Science Pole, which put the average age at less than forty. If anything, the increase in younger workers was a growing trend from the later 1980s into the mid 1990s. See W. Meske and F. de Alaiza, *Scientiometrics* 18, nos. 1–2 (1990): 137–55. My thanks to one of the reviewers for this clarification.

19. Finlay Institute brochure, "*Más de cien años de tradición científica*" (n.d.; likely 2000–2001); see also L. Margulis and T. H. Kunz, "Glimpses of Biomedical Research and Education in Cuba," *Bioscience* 34, no. 10 (1984): 634.
20. Doreen Massey, Paul Quintas, and David Wield, *High-Tech Fantasies: Science Parks in Society, Science and Space* (Routledge: London, 1992), 43.
21. *Nauchnye kadry Leningrada*, 106–12, quoted in P. Kneen, *Soviet Scientists and the State: An Examination of the Social and Political Aspects of Science in the USSR* (Hong Kong: Macmillan, 1984).
22. In the 1980s, it was common for Cuban scientists to train abroad, in countries that included Germany, Argentina, Belgium, Brazil, Czechoslovakia, Spain, Finland, Hungary, Italy, Japan, the Low Countries, the United Kingdom, Sweden, Switzerland, and the USSR, for example. As we have seen, some had also trained at laboratories in the United States.
23. *¿Quien es quien en las biociencias cubanas?*, CD-ROM produced and supplied by the Ministry for Science, Technology, and the Environment (CITMA).
24. David Dickson, *The New Politics of Science* (Chicago: University of Chicago Press, 1998).
25. G. Trueba-Gonzáles, *Política científica y biotecnología en Cuba: 5a curso de planificación de ciencia y tecnología en América Latina* (Havana, 1991), 8. Trueba-González is director of the Institute of Economic Research at JUCEPLAN (Central Planning Board).
26. Steven Shapin, *A Social History of Truth: Civility and Science in Seventeenth-Century England* (London: University of Chicago Press, 1994), 301.
27. Ibid., 302.
28. Accord No. 1157, Biblioteca Nacional José Martí (henceforth BNJM).
29. Gustavo Sierra, interview, May 8, 2002.
30. Ernesto Bravo, *Development within Underdevelopment: New Trends in Cuban Medicine* (Havana: Editorial José Martí/Elfos Scientiae, 1998), 135.
31. On the Lysenko affair and its broader implications, see D. Joravsky, *The Lysenko Affair* (Cambridge, MA: Harvard University Press, 1970); D. Lecourt, *Proletarian Science? The Case of Lysenko* (London: New Left Books, 1977); and G. Loren, *Science, Philosophy, and Human Behavior in the Soviet Union* (New York: Columbia University Press, 1987).
32. Kneen, *Soviet Scientists and the State*.
33. A UNIDO visitor to Cuba commented, a little simply, that "life is austere for the Cubans, but, at least for the scientists, it is bracing and stimulating," *Science* 169 (July 1970): 34.
34. J. Richardson, "Research in Cuba Today," *Science and Public Policy* 21, no. 3 (1994): 185–86.
35. This is not to say similar informalization was not present in the former Soviet Union. As Kneen shows in *Soviet Scientists and the State*, a study carried out by the Institute of Molecular Biology of the Soviet Academy of Sciences on informal interlaboratory relations from the institute's founding in 1959 through 1977 found that "the informal arrangements tended to outgrow the formal framework. This occurred with the burgeoning of personal communications between the members of thematic groups, into which laboratories are subdivided, which could be observed for a year or two prior to the transformation of these groups into independent, officially recognized units. . . . The formal structure and the official goals of the institute could thus be seen to respond to the changing informal relations of scientists."

36. Robert Putnam, *Bowling Alone: The Collapse and Revival of American Community* (New York: Simon & Schuster, 2000).
37. Mark Granovetter, "Economic Action and Social Structure: The Problem of Embeddedness," *American Journal of Sociology* 91, no. 3 (1985): 481–510.
38. *Granma*, July 14, 1986.
39. Diego Gambetta, *Trust: Making and Breaking Cooperative Relations* (Oxford: Basil Blackwell, 1990). See also Piotr Stzompka, *Trust: A Sociological Theory* (Cambridge: Cambridge University Press, 1999).
40. Harvey Bialy, the editor of *Bio/Technology*, concurs that the Cubans work "under extreme pressures" ("Biotechnology and Economic Development: Introduction," *Economic Bulletin for Europe* [journal of the United Nations Economic Commission for Europe] 38, no. 1 [March 1986]: 7).
41. Saxenian, *Regional Advantage*, 38.
42. Antoni Kapcia, *Cuba: Island of Dreams* (Oxford: Berg, 2000).
43. Paul Rabinow, *Making PCR: A Story of Biotechnology* (Chicago: University of Chicago Press, 1996), 164; Paul Wilberg, *Deep Socialism: A New Socialist Manifesto* (London: Third Ear Limited, 1998), makes a similar point.
44. This according to its former head: I. M. Smith, *Report on NACSEX 1/28–2/7/90 Trip* (Havana, 1990).
45. Fidel Castro, FBIS: 1992-062, available at http://lanic.utexas.edu/la/cb/cuba/castro.html.
46. One prominent scientist noted the "resentment among the less well-endowed research institutes and university faculties, particularly because resources in general are scarce": Smith, *Report on NACSEX*.
47. It was "necessary to introduce mercantilist monetary relations into scientific and technological activity which would contribute to obtaining satisfactory results"; in *"Tercer por cuanto la revolución,"* Conjunta CEF-ACC, December 9, 1985, 107, BNJM.
48. Comité Estatal de Finanzas, Resolución 2/85: "en las empresas en que se obtengan resultados económicos favorables por la aplicación de innovaciones o racionalizaciones, se cree un fondo con el importe del veinte porciento de dicho resultado económico favourable, 7 enero 1985," *Gaceta Oficial* 83, no. 4 (March 4, 1985): 66–67.
49. Hence, Resolution 67/85, which demanded the development of a system of financial norms and forms of accounting, did not apply to these centers: "Resolución 67–85 del Ministro-Presidente del Comité Estatal de Finanzas por la que se pone en vigor el Procedimiento para la Elaboración, Cálculo y Determinación de las Normas Financieras de los Modelos de Rotación de las Unidades de Ciencia y técnica que operan bajo el sistema de cálculo económico, 19 diciembre 1985," *Gaceta Oficial* 83, no. 89 (December 23, 1985) : 1473–74.
50. Cuban Academy of Sciences, *El Estado Actual*, 1981, BNJM.
51. Olga Miranda, "Financiamiento de la investigación en genética en America Latina. Genetica: V congreso LatinoAmericano," *Asociación LatinoAmericano de Genética, Sociedad de Genetica de Chile* (Santiago, 1982).
52. Herbert Marcuse, *One Dimensional Man*, 2nd ed. (Boston: Beacon Press, 1991).
53. John Dewey, *Essays in Experimental Logic* (New York: Dover Books, 1953).
54. ExpoCuba, 1989, 12.
55. Joy Gordon, "Cuba's Entrepreneurial Socialism," *Atlantic Monthly* 279, no. 1 (1997).

CHAPTER FIVE
1. Agustín Lage, ABC News broadcast, 2001.
2. *Cuba Business*, 1990, n.d.
3. Douglas Starr, "The Cuban Biotech Revolution," *Wired* magazine, December 12, 2004.
4. Measured from the level it had achieved in 1989. By 1993 the fiscal deficit reached 33 percent of GDP (*Economist* [2001]: 18).
5. Fidel Castro, Foreign Broadcast Information Service (henceforth FBIS), 1991–199; Castro's speeches are documented at the FBIS Castro Speech Database, available at http://lanic.utexas.edu/la/cb/cuba/castro.html.
6. For example, Articles 14, 15, and 18 of the Cuban constitution regarding ownership of property were adapted.
7. Fidel Castro, FBIS: 2002-04-91, available at http://lanic.utexas.edu/la/cb/cuba/castro.html.
8. The number of foreign and Cuban firms attending the fairs rose from 400 in 1988, to 816 in 1991, and again to 3,000 in 1996. Source: *Business TIPS on Cuba*, data drawn from various years.
9. J. Benjamin-Alvarado, *Non-issue: Cuba's Mothballed Nuclear Power Plant* (Washington, DC: Center for International Policy, 2003).
10. By 1994 total expenditure on science and technology was down to 88 percent that of 1990. Expenditure did not surpass the 1990 level again until 1998. The halt in expansion is recorded in the annual expenditure on physical infrastructure which was 56 percent of the 1990 level in 1994, and on the whole continued to decline throughout the decade while, despite an even more severe initial drop to 29 percent of 1990 levels by 1994, expenditure on capital improvements (such as new computer equipment and biosafety equipment) investment in this area gradually increased. Source: *Cuba en Cifras*, OFNE, 1999: table 16.5, n.p.
11. "Made in Cuba," *Time* magazine 147, no. 20 (May 13, 1996). With this in mind, initial but not wholly successful attempts were made to repeat the experience of the Science Pole in Camagüey and Sancti Spíritus in the far east of the country (though for a discussion of the work of this center, see J. A. Gómez and A. Lima, "El orígen y desarrollo de la biotecnología en la província de Cienfuegos," *Revista Finlay* 5, no. 1 (1991): 96–101. My focus here remains on the Havana Science Pole though.
12. A summary account of PPG can be found in Sheldon S. Hendler, "A Cuban Issue," *Journal of Medicinal Food* 4, no. 2 (June 2001): 55–56.
13. Adele E. Clarke, Janet K. Shim, Laura Mamo, Jennifer Ruth Fosket, and Jennifer R. Fishman, "Biomedicalization: Technoscientific Transformations of Health, Illness, and U.S. Biomedicine," *American Sociological Review* 68, no. 2 (April 2003): 161–94.
14. Different takes on these different histories are provided by Armin C. Braun, *The Cancer Problem* (New York: Columbia University Press, 1969); Lelland J. Rather, *The Genesis of Cancer: A Study in the History of Ideas* (Baltimore, MD: The Johns Hopkins University Press, 1978); James T. Patterson, *The Dread Disease* (Cambridge, MA: Harvard University Press, 1987); and Robert Proctor, *Cancer Wars* (New York: Basic Books, 1995).
15. The current great interest in this approach did not begin until the mid 1980s and only really blossomed in the early 1990s: "Whether the immunology community has believed cancer immunotherapy is feasible or impossible has been largely determined by the prevailing immunological paradigms. In fact, during the last 110 years it is possible to trace at least five dramatic fluctuations in attitude toward cancer im-

munotherapy." See C. R. Parish, "Cancer Immunotherapy: The Past, the Present and the Future," *Immunology and Cell Biology* 81 (2003): 106–13.

16. Cornelius Rhoads, the director of the Sloan-Kettering Research Institute in Manhattan, cited in R. F. Budd, "Strategy in American Cancer Research post World War Two," *Social Studies of Science* 8, no. 4 (November 1978): 425–59), quotation at 434. Rhoads was writing to a contemporary on the subject of goal definition in science and specifically in relation to his colleague Kettering's style of work.
17. Unnamed molecular biologist, cited in Joan Fujimura, *Crafting Science: A Sociohistory of the Quest for the Genetics of Cancer* (Cambridge, MA: Harvard University Press, 1996), 184.
18. Ibid.
19. A. A. Martin, Y. H. Galan, A. J. Rodriguez, M. Graupera, et al., "The Cuban National Cancer Registry, 1986–1990," *European Journal of Epidemiology* 14, no. 3 (April 1998): 287–97.
20. Global cancer rates could increase by 50 percent to 15 million by 2020; see www.who.int/mediacentre/news/releases/2003/pr27/en/ (based on that year's World Cancer Report).
21. Ian Davis, "An Overview of Cancer Immunotherapy," *Immunology and Cell Biology* 78 (2000): 179–95: quotation at 180. See also Stephen S. Hall's *A Commotion in the Blood* (1997), which describes immunotherapy as a "quiet revolution" in medicine.
22. Georges Kohler and César Milstein, "Continuous Cultures of Fused Cells Secreting Antibody of Predefined Specificity," *Nature* 256 (1975): 495–97.
23. Compare Budd, "Strategy in American Cancer Research." This is not to say there are not important affinities between the industrial style of cancer research promoted in the United States in the immediate postwar period and in Cuba.
24. R. Pérez, M. Pascual, A. Macías, and A. Lage, "Epidermal Growth Factor Receptors in Human Breast Cancer," *Breast Cancer Research and Treatment* 4, no. 3 (1984): 189–93.
25. See Paul Harari, "Epidermal Growth Factor Receptor Inhibition Strategies in Oncology," *Endocrine-Related Cancer* 11 (1994): 689–708, 690.
26. This work continued some of what Pérez had been doing in Nice. See R. Pérez, J. C. Chambard, E. Van Obberghen-Schilling, A. Franchi, and J. Pouyssegur, "Emergence of Hamster Fibroblast Tumors in Nude Mice: Evidence for In Vivo Selection Leading to Loss of Growth Factor Requirement," *Journal of Cellular Physiology* 109 (1981): 387–96, and R. Pérez, A. Franchi, B. F. Deys, and J. Pouyssegur, "Evidence That Hamster Fibroblast Tumors Emerge in Nude Mice through the Process of Two In Vivo Selections Leading to Growth Factor Relaxation and Immune Resistance," *International Journal of Cancer* 29 (1982): 309–14.
27. Jorge Lombardero, Rolando Pérez, and Agustín Lage, "Epidermal Growth Factor Inhibits Thymidine Incorporation in Erhlich Ascites Tumor Cells In Vivo," *Neoplasma* 33, no. 4 (1986): 423–29.
28. The editor of the journal *Bio/Technology* commented, after a brief visit to Cuba, on the innovative approach being taken by the Cubans at INOR to cancer research (Harvey Bialy, 1988: n.p.).
29. For example, José Baselga and John Mendelsohn, working at Sloan-Kettering in New York; Chaitanya Divgi et al. also working at Sloan-Kettering, and Luther Brady et al. at Hahnemann University in Philadelphia.
30. Ian D. Davis, Michael Jefford, Philip Parente, and Jonathan Cebon, "Rational ap-

31. proaches to human cancer immunotherapy," *Journal of Leukocyte Biology* 73 (2003): 3–29. See also Davis (2000).
31. G. González et al., "Induction of Immune Recognition of Self Epidermal Growth Factor (EGF): Effect on EGF Biodistribution and Tumor Growth," *Vaccine Research* 5 (1996): 233–44.
32. Agustín Lage, "On the Cross-Fertilization between Biotechnology and Immunology: Current Situation in Cuba," *Vaccine* 24, no. 12, suppl. 2 (2006): S3–6.
33. John Gribben, Cancer Research UK's Medical Oncology Unit, telephone interview, London, January 11, 2007. Gribben went on to point out that, in the United Kingdom, even to do animal experiments was difficult with animal activists present. To then further this work into early clinical trials also required jumping a vast number of regulatory hurdles, followed by the economic costs of later-stage clinical trials which had to be in concert with a pharmaceutical company. At the end of the day, some drugs wouldn't reach patients, he suggested, because they proved not to be "doable."
34. York Medical later changed its name to YM Biosciences. In the text here, York Medical refers to the period up until it changed its name.
35. YM Biosciences, *Annual Report* (2001): 9.
36. "Equity4life Bets Its Investments Will Shape Tomorrow's Medicine," *Bloomberg Report* (December 14, 2000), 1.
37. David Harrington, "The Randomized Clinical Trial," *Journal of the American Statistical Association* 95, no. 449 (2000): 312–15.
38. David Allan, interview, August 20, 2002.
39. Jack Kincaid, former CEO, York Medical, quoted in "Cuba Investment Report" (Cuba: Havana Asset Management, December 1997).
40. "Biociencias en Cuba: Actualización," IDICT Biomundi Consultoría (1995): 25.
41. "La Economía Cubana 1995 y 1996," Oficina Nacional de Estadísticas (OFNE) (Havana, 1996): 5.
42. "Company Profile: York Medical," HAM (Havana, December 1997): 1.
43. Reported in www.globetechnology.com, "Biotech Builds on Cuban Innovation," May 2, 2001.
44. "Law on Investments Consolidates Economic Opening," *Business TIPS on Cuba* (January 1996): 60.
45. PCC, "Resolución económica del V congreso del PCC," *Granma Suplemento Especial*, October 13, 1997.
46. Ernesto Grillo, interview with *Avances Médicos de Cuba* 4, no. 10 (1997): 37.
47. "Decree of the Minister for Science, Technology, and the Environment," 1997 [original in English], personal copy.
48. American Association for the Advancement of Science, "A Window on Cuba's Biotech Initiative," internal document, personal copy.
49. *Clinica* 798 (March 2, 1998): n.p.
50. Abelardo Márquez, quoted in *Avances Médicas de Cuba* 4, no. 10 (1997): 37.
51. This is evident in the report "Selección de compañías como posibles socios en el sector Vacunas," Biomundi (1998).
52. YM Biosciences, *Annual Report* (2001): 19.
53. Susan Star and James Griesemer, "Institutional Ecology," *Social Studies of Science* 19 (1989): 387–420. See also Thomas Gieryn, "Boundary-Work and the Demarcation of Science from Non-Science: Strains and Interests in Professional Ideologies of

Scientists," *American Sociological Review* 48 (1983): 781–95. Fujimura also talks of "boundary objects" (see her chapter in *Science as Practice and Culture*, ed. Andrew Pickering (Chicago: University of Chicago Press, 1992).

54. David Allan, interview, August 26, 2002.
55. In a paper written nearly ten years later, Lage himself even confessed to finding it hard to name what it was that made Cuban research different: Agustín Lage, "On the Cross-Fertilization between Biotechnology and Iimmunology: Current Situation in Cuba," *Vaccine* 24, no. 12, suppl. 2 (2006): S3–6.
56. François Delaporte, *The History of Yellow Fever: An Essay on the Birth of Tropical Medicine* (Cambridge, MA: MIT Press, 1991). The Nietzche quote on this page is also from Delaporte, *The History of Yellow Fever*.

CHAPTER SIX

1. Specifically Rogés referred here to taking representatives from CECMED and CENCEC to Canada's Health Protection Board—the equivalent medical quality control institution there.
2. Biognosis Limited, BioPortfolio, available at http://www.bioportfolio.com/biocorporate/1563-Biognosis+Limited.html, accessed September 16, 2009.
3. As described in James Rossiter, "Kingsnorth Takes Control of Beta Gran Caribe," *efinancial news*, October 10, 2001. The article can be accessed at http://www.cubatrade.org/2003Lst.pdf.
4. See, e.g., the article by Gareth Jenkins, then head of the London-based company Cuba Business and closely acquainted with the issue of Cuban investments, "Investment Funds in Cuba: An Upcoming Caribbean Tiger," *Colombia Journal of World Business* (Spring 1995): 43–48.
5. See Markus Nolff, *TRIPs, PCT, and Global Patent Procurement* (London: Kluwer Law International, 2000).
6. See, e.g., Peter Drahos and Ruth Mayne, *Global Intellectual Property Rights: Knowledge, Access and Development* (London: Palgrave Macmillan, 2003), and Bronwyn Parry, "Cultures of Knowledge Production: Investigating Intellectual Property Rights and Relations in the Pacific," *Antipode* 34, no. 4 (2002): 679–707.
7. David Bainbridge, *Intellectual Property* (London: Prentice Hall, 1996), 265. Of course, patents are not unproblematic indicators of invention or technological output, and invention in one sector may equate to a patent in another (see Organisation for Economic Co-operation and Development, *Science, Technology and Industry: Scoreboard of Indicators* [Paris: OECD, 1997]). Here, however, I am more interested in the strategic use of patents rather than their value as an indicator of innovation; indeed, the value of such a use is problematic for the very reasons outlined in this chapter. See also Adronico Adede, "The Political Economy of the TRIPS Agreement: Origins and History of Negotiations" (2001), at www.ictsd.org/dlogue/2001-07-30/Adede.pdf.
8. For a critique of Western intellectual-property rights regimes, see Grant Hammond, "The Legal Protection of Ideas, Part 1," *Computer Law and Security Report* 8, no. 3 (May–June 1992) 102–10. On the GATT-TRIPS, see Carlos Correa, *Intellectual Property Rights, the WTO and Developing Countries: The TRIPS Agreement and Policy Options* (London: Zed Books, 2000). And for a geographical account of the GATT-TRIPS as an emergent regulatory system, see B. Parry, *Trading the Genome* (New York: Columbia University Press, 2004), 78–80. The signatory countries were Argentina, Brazil, Chile, China, Colombia, Cuba, Egypt, India, Nigeria, Peru, Tanzania, and Uruguay,

later joined by Pakistan and Zimbabwe. For examples of some of the proposals mentioned, see "Submission by the African Group" (IP/C/W/296, June 29, 2001) on TRIPS and Public Health, and the "Draft Ministerial Declaration" (IP/C/W/312/WT/GC/W/450, October 4, 2001) presented to the General Council of the WTO by a group of countries including Cuba demanding additional leniency in meeting TRIPS requirements owing to their special circumstances.

9. Jacques Gorlin, *An Analysis of the Pharmaceutical-Related Provisions of the WTO TRIPS (Intellectual Property) Agreement* (London: Intellectual Property Institute, 1994), 1. See also Michael E. Blakeney, *Border Control of Intellectual Property Rights* (London: Sweet & Maxwell, 2001), v.

10. Rosemary Coombe, "Authorial Cartographies: Mapping Proprietary Borders in a Less-Than-Brave New World," *Stanford Law Review* 48 (1996): 1357–66. See also Rosemary Coombe, *The Cultural Life of Intellectual Properties* (Durham, NC: Duke University Press, 1998).

11. Perhaps the classic example of this "bargaining" is Article 65.4 of the TRIPS Agreement in which most developing country members are provided with ten-year (January 1, 1995–January 1, 2005) delayed application of patent protection for pharmaceutical products. See Marney L. Cheek, "The Limits of Informal Regulatory Cooperation in International Affairs: A Review of the Global Intellectual Property Regime," *George Washington Journal of International Law Review* 33 (2001): 277–323.

12. American Declaration of the Rights and Duties of Man, Article 23.

13. Resolution 28, September 10, 1973, Dr. Carlos Rafael Rodríguez, vice president of the Council of Ministers and ONIITEM, "Objetivos y principales funciones de la ONIITEM" (1982): 27.

14. ONIITEM, "La información de patentes: motor impulsor de la actividad creadora" (1984). See also Camara Comercial de Cuba, "La Propiedad Industrial en Cuba" (n.d.).

15. On how the patent was to be worked, see in particular Article 48, which prevents Cuban citizens from registering any patents abroad that have not been first registered in Cuba, and Article 66, which, as is common practice elsewhere, grants the right to work a patent exclusively to the owners of the patent, not the authors of the invention, but more strongly than elsewhere; the state (via the national intellectual property office) has the right to compulsory license any such patents not also being worked by those institutional owners of the patents, such as the research institutes of the Science Pole. Such legislation still allowed, via Article 63, for the importation of patented products not registered in Cuba, effectively still permitting the practice of reverse engineering. At the same time, Article 104 specified for scientific inventions that any remuneration would only be granted while the inventor was a resident of Cuba. The effect of these various articles is to bind the whole patenting process much more strongly to the central state itself, and to otherwise more strongly nationalize the process, than is the norm in the United States or Europe. This is affirmed and specified in Articles 84, 85, and 87. Ultimately even the invention itself, and not just the right to work it, could be argued to be the fundamental property of the state.

16. ONIITEM (1982): 12; emphasis added.

17. This is set out more clearly in Resolución 999/83 (1983), which accompanied Decree Law No. 68 (1983). See, in particular, Articles 4, 14, and 20.

18. I borrow this phrase from Kathleen McAfee, "Neoliberalism on the Molecular

Scale: Economic and Political Reductionism in Biotechnology Battles," *Geoforum* 34 (2003): 203–19.
19. "La Propiedad Industrial en Cuba," Camara de Comercio de Cuba (n.d.).
20. *Granma*, November 11, 1999.
21. There are many more international agreements, and I do not consider here those relating solely to agricultural or plant biotechnology, such as the International Undertaking (IU)—a legally nonbinding instrument of the Food and Agriculture Organization (FAO) of the United Nations to regulate trade in seeds—or the Plant Genetic Resources Agreement, now the Commission on Genetic Resources for Food and Agriculture (CGRFA). The interrelationships between these various (sometimes overlapping and sometimes contradictory) agreements are complex. They are considered here only in so far as they relate to the situation in Cuba.
22. Resolución no. 66/96 de 15 Julio de 1996 de la Ministra de Ciencia, Tecnología y Medio Ambiente de la República de Cuba, "Normas para la Aplicación del Tratado de Cooperacion en Materia de Patentes (PCT) en la República de Cuba." Cuba was a signatory to the Paris Agreement (it joined in 1904), WIPO (it joined in 1975), the PCT (it joined in 1996), and the Treaty of Budapest (it joined in 1994). By the late 1990s it was a member of seventeen international IP treaties in total.
23. Biomundi, "Breve perfil de varios medicamentos cubanos: Informacíon general acerca de los procesos para el registro de patentes y el registro de medicamentos en Cuba" (August, 1998).
24. Decree Law No. 160, Consejo de Estado.
25. Rosa Elena Simeón, minister for Science, Technology, and the Environment, cited in *Cuba Business* 10, no. 7 (1996): 3.
26. Rose Elena Simeón, minister for Science, Technology, and the Environment, cited in *Granma*, December 30, 1995.
27. Cuba, "Decreto Ley No. 160 para facilitar la presentación y modificación de solicitudes de patentes para productos farmaceúticos y químicos para la agricultura," Article 3.3, 2.
28. The quote on patenting as a sociocultural paradigm comes from Joan Fujimura, *Crafting Science: A Sociohistory of the Quest for the Genetics of Cancer* (Cambridge, MA: Harvard University Press, 1997), 18; the latter from Bainbridge, *Intellectual Property*, 157.
29. See Jesús Sanchelima, "Selected Aspects of Cuba's Intellectual Property Laws," in *Cuba in Transition*, Papers and Proceedings of the Twelfth Annual Meeting of the Association for the Study of the Cuban Economy (ASCE) (2002), 12:215.
30. Luis Enrique Fernández, Vaccines Division director at the CIM, in an interview with *Medicc Review*, Havana, September 2004.
31. These figures are from Michelle Frank and Gail Reed, *Denial of Food and Medicine: The Impact of the U.S. Embargo on Health and Nutrition in Cuba*, American Association for World Health (1997: 35). Available online at http://www.usaengage.org/studies/cuba.html.
32. Helms-Burton Act, Title 3§301(5). Such extraterritoriality has been heavily criticized by legal experts. See, e.g., Robert Muse, "A Public International Law Critique of the Extraterritorial Jurisdiction of the Helms-Burton Act (Cuban Liberty and Democratic Solidarity [LIBERTAD] Act) of 1996," *George Washington University Journal of International Law and Economics* 30, nos. 2–3 (1996–97): 207–70; and M. A. Groombridge, "Missing the Target: The Failure of the Helms-Burton Act" (Washington: CATO Institute, 2001): 12. The European Union has also complained that the

Helms-Burton is in breach of Articles 1, 2, 5, and 11 of the General Agreement on Tariffs and Trade (GATT). Others have sought to defend it, of course: David P. Fidler, "LIBERTAD v. Liberalism: An Analysis of the Helms-Burton Act from within Liberal International Relations Theory," *Indiana Journal of Global Legal Studies* 297, no. 4 (2002): 297–355.

33. Donna Rich Kaplowitz, *Anatomy of a Failed Embargo* (Colorado: Lynne Reimer Publishers, 1998). See, e.g., the anonymous Cuban pamphlet, *"Aqui no queremos amos— Here We Want No Masters"* (Havana: Editora Politica, 1998).

34. "Viola primera enmienda constitucional la prohibición de articulos cientificos de Cuba, Irán, Libia y Sudan," *Granma Internacional*, February 29, 2004.

35. Vice Minister Ramón Díaz-Vallina, MINSAP, cited in *Cuba Business* 10, no. 4 (1996): 3.

36. See http://www.in-cites.com/papers/dr-salvador-moncada.html.

37. "Cuba's Entrepreneurial Socialism," *Atlantic Monthly* (January 1997): 18.

38. Joseph Camilleri and Jim Falk, *The End of Sovereignty? The Politics of a Shrinking and Fragmenting World* (Aldershot: Edward Elgar, 1992).

39. Stephen Gill, "European Governance and New Constitutionalism: Economic and Monetary Union and Alternatives to Disciplinary Neoliberalism in Europe," *New Political Economy* 3 (1998): 5–26.

40. Rosemary Coombe, "The Cultural Life of Things: Anthropological Approaches to Law and Society in Conditions of Globalisation," *University Journal of International Law and Policy* 10 (1995), 828. Another critical legal scholar, Keith Aoki, makes a similar point when discussing digital information. He talks of "sharp disputes over intellectual property protection [that] arise between culturally, linguistically, and physically distant nations, such as the United States and China" ("[Intellectual] Property and Sovereignty: Notes toward a Cultural Geography of Authorship," *Stanford Law Review* 48 [1996]: 1293–1305).

41. Kevin Burch, "The 'Properties' of the State System and Global Capitalism," in *The Global Economy as Political Space*, ed. S. Rosow, N. Inayatullah, and M. Rupert (London: Lynne Rienner, 1994), 37–59.

42. See Alan Hudson, "Offshoreness, Globalization and Sovereignty: A Postmodern Geo-Political Economy?" *Transactions of the Institute of British Geographers* 25, no. 3 (2000): 269–83, who makes use of a similar notion of "regulatory skirmishes." I am indebted to Alan's work and many conversations for sparking the ideas I develop here.

43. R. Weissmann, "A Long, Strange TRIPS: The Pharmaceutical Industry Drive To Harmonize Global Intellectual Property Rules and the Remaining WTO Alternatives Available to Third World Countries," *University of Pennsylvania Journal of International Economic Law* 17 (1996): 1069–1125.

44. http://www.cigb.edu/cigb_e.html.

45. Carlos Mella, CML, vice director, Heber Biotec, interview, July 10, 2002.

46. This was completed with the establishment of a further two centers: the Center for State Control of Medical Equipment in 1992 and the Center for the Development of Pharmaco-epidemiology in 1996. A new national policy incorporating all these bodies was then outlined in 2000.

47. *Business TIPS on Cuba* (June 1994), 22.

48. For an overview of the initial aims of CECMED, see R. Darias Rodés, "Aseguramiento de la calidad en biotecnología," *Normalización* 21, no. 2 (1992): 3–27.

49. A good account of the emergence and operation of contract research organizations

is provided in Adriana Petryna, "Globalizing Human Subjects Research," in *Global Pharmaceuticals: Ethics, Markets, Practices*, ed. Adriana Petryna, Andrew Lakoff, and Arthur Kleinman (Durham, NC: Duke University Press, 2006), 33–60, esp. 38.

50. Accordingly, CECMED and CENCEC were part of the new two-tier National Science and Technology System (NSTS) in Cuba: NSTS, "Improvement of the Science and Technology System in Cuba," personal copy.

51. See "Información en Propiedad Industrial," OCPI, http://www.ocpi.cu/propiedad00.html.

52. These being: Bufete Internacional SA, Conas SA/Consultores Asociados, SA, Consultoría Jurídica Internacional, SA, Claim SA, and Lex SA.

53. Eduardo Orozco, the director of Biomundi, commented to me: "I would imagine that there are around ten biotech people who are deputies of the national assembly. Biotechnology is the scientific sector that enjoys the greatest representation in the National Assembly" (Havana, 2003).

54. See, respectively, *Biociencias en Cuba*, Biomundi (various years); *Bioglosario*, Biomundi (various years); *Guia Biomundi para la exportación a Cuba*, Biomundi (various years); *Quién es quién en la biociencias cubana*, CD-ROM; *Turismo de negocios Biomundi: Una oferta para descansar con beneficios*, Biomundi (1996); *Biociencias en Cuba: Actualización*, IDICT Biomundi Consultoría (Havana: IDICT Publishers, 1995), 25.

55. Lesley Munro-Faure, "Achieving the New International Quality Standards," *Transatlantic Publications*, 1995; Francis Buttle, "An Investigation of the Willingness of UK Certificated Firms to Recommend ISO 9000," 1996. For further background to the ISO, see http://www.iso.org/iso/en/is09000-14000/index.html.

56. Ian Nicholls, *ISO 9000: The Key to International Markets* (Bewdley: Nicholls & Nicholls, 1993).

57. Reports NC 92–01–1: 91, Comité Estatal de Normalización, Havana. Interestingly, in the elaboration of several quality control studies reviewed, Japanese texts were cited. While this may be due to availability, it may also reflect a certain interest in the Japanese model over a more obviously Euro-American model evidenced in several interviews.

58. O. Y. Freyre, "Enfoque actual para elevar la calidad," *Normalización* 21, no. 2 (1991): 25.

59. Report NC 26–212 (1992), Comité Estatal de Normalización, Havana.

60. See, respectively, Regulación no. 2–95; Regulación no. 11–98; Regulación no. 14–98; and Regulación no. 18–99.

61. E. C. Batíz, "Modelo de gestion de la seguridad y la bioseguridad en centros de la biotecnología," Facultad de Ingeniaría Industrial (Havana: Instituto Superior Politécnico "José A. Echeverría," 1996), 2.

62. Reports NC-26-212: (1992) and NC-26-211 (1992), respectively, Comité Estatal de Normalización, Havana.

63. This phrase was used in a report by L. Gómez-Napier, a titular researcher at the CIGB (*Diseño y aplicación de un sistema de la calidad en las producciónes biotecnológicas* [Havana: CIGB, 1995], 1).

64. This was noted, e.g., in S. Viña, E. Concepción Batís, and R. Montero Martínez, "Experiencias en la integración de la calidad y la seguridad en la industría biotecnológica y farmaceútica," *Ingeniería Industrial*, n.d.

65. See, e.g., CECMED, "*Requisitos para las solicitudes de inscripción, renovación y modificación en el registro de medicamentos de uso humano*" (1998).

66. This quote is taken from an interview between José de la Fuente and Jens Plahte, a scholar at Norway's Center for Technology, Innovation, and Culture, conducted in

Oklahoma, March 2002. I am extremely grateful to Jens for sharing some of his own research into Cuban biotechnology with me.
67. *United States of America v. Sam Waksal*, indictment. Waksal was ultimately sentenced to seven years in prison. See also David Usborne, "ImClone's Waksal Jailed for Seven Years," London *Independent*, June 11, 2003.

CHAPTER SEVEN
1. Jocelyn Kaiser, *Science* 282 (1998): 1626.
2. Alejandro Gonzáles, government spokesman quoted by Reuters, June 24, 2009.
3. *Business TIPS on Cuba*, April 1999, 12.
4. "Focus on Biotech," *Cuba Business* (June 1997): 7.
5. Agustín Lage, quoted by Reuters, Havana, April 14, 1999.
6. Patricia Zengerle, "Firm Wants To Work with Cuba on Meningitis Vaccine," Reuters, June 5, 1998; emphasis added.
7. *Cuba Business* 13, no. 6 (July–August 1999): n.p.
8. "U.S. Finally Will Let SmithKline Market Cuban Meningitis Vaccine," *Wall Street Journal*, July 23, 1999.
9. "Breakthrough for Cuban Biotech," *Science* 285, no. 5434 (September 10, 1999): 1663.
10. "Focus on Biotech," *Cuba Business* (June 1999): 7.
11. Fidel Castro, "Speech at the Summit Meeting of Latin America, the Caribbean and the European Union" [original in English].
12. See, e.g., "Finding Ways to Dabble in Cuba, Legally," *New York Times*, March 5, 2000.
13. Small molecules comprise a diverse group of natural and synthetic substances which can be targeted specifically toward cancer cells.
14. The first was Panorex, developed by Centocor Inc. and approved in Germany in 1994, and the second was Herceptin, developed by Genentech and approved in the United States in 1998.
15. According to a CIM press release, March 2001, personal copy.
16. US-Cuba Trade and Economic Council, *Economic Eye on Cuba*, September 2000, personal copy.
17. "Biovation and Centre of Molecular Immunology Cuba announce Patent Licence Agreement," CIM press release, October 24, 2000.
18. "Roundup Ready Cigars?" www.americas.org/news/features/200007_biotechnology/cuba_monsanto.asp.
19. Agustín Lage, "Las biotecnologías y la nueva economía: Crear y valorizar los bienes intangibles," *Biotecnología Aplicada* 17 (2000): 55–61.
20. Ibid., 55.
21. Indeed, articles debating the pros and cons of various forms of intellectual property treaty and their implications for Cuba were still appearing in 2000 (e.g., Jorge Espinosa, 2000).
22. John Law, 25.
23. *Avances Medicos de Cuba* 1, no. 1 (1994): 2.
24. *Avances Medicos de Cuba* 4, no. 10 (1997): 25.
25. This was mentioned at a Parliamentary meeting on Cuban biotech in London that I attended at Portcullis House on March 26, 2004.
26. In 2002 I conducted a biotech industry alliance survey that aimed to find out what Western biotech directors are looking for when considering alliances or mergers.

This sort of survey has been carried out before—it is very much stock-in-trade for accounting firms, such as PriceWaterhouseCoopers, that produce regular overviews of the pharmaceutical industry. My much more modest survey was focused only in those areas the Cubans work, and it was sent only to companies that, in their corporate strategy, espouse an interest in North-South alliances. Under investigation was not whether they might consider an alliance with places like Cuba, therefore, but on what terms. The level of responses does not permit generalizations to be drawn. But what many of the responses do give a flavor of is the mindset that is taken to countries such as Cuba; through this we can gain a sense of how geographical misunderstandings help to fashion a more directed vocabulary of suspicion. In short, while tolerating many forms of uncertainty in their own laboratory and business practices, they set the barriers somewhat higher for potential South biotech partners. The tone of response indicated that many North companies would consider South alliances only on their own terms.

27. This of course was a reworking of the personal body/body politic trope in Cuban political discourse more generally: see Antonio Benítez-Rojo, *The Repeating Island: The Caribbean and the Postmodern Perspective* (London: Duke University Press, 1992), and Damian Fernandez, *Cuba and the Politics of Passion* (Austin: University of Texas Press, 2000).

28. José de la Fuente, "Wine into Vinegar: The Fall of Cuba's Biotechnology," *Nature Biotechnology* 19 (2001): 905.

29. Few people were willing to discuss de la Fuente's article with me directly or, for that matter, any of the events that lead up to his exit from the CIGB and the country. At the same time, top officials were removed from their posts at the Hermanos Ameijeiras Hospital in central Havana, Cuba's principal biomedical research hospital, in that instance on grounds of mishandling of funds and as part of an explicitly stated campaign against "antisocial" activities. Like the biotechnology centers, the hospital is run by the State Council and not MINSAP, which is in charge of the other hospitals. The *Miami Herald*, e.g., carried the headline "Havana Hospital Officials Forced Out in Corruption Crackdown," on July 2, 1999. The Hermanos Ameijeiras Hospital has strong links with the CIGB and is used for much of its clinical testing.

30. Julián Alvarez Blanco, "Letter to the Editor," *Nature* (October 19, 2001), personal copy.

31. Source: OFNE, 1999, table 6.12, "Exportaciones," 141.

32. US Department of State, "Beyond the Axis of Evil: Additional Threats from Weapons of Mass Destruction," John R. Bolton, Undersecretary for Arms Control and International Security, speech to the Heritage Foundation, Washington DC, May 6, 2002. Cuba had signed a deal to transfer hepatitis B production technology to Iran when Castro visited the country in May 2001 (*Iran Report* 4, no. 29 [August 6, 2001]). Certain of the impetus for these claims came from Miami, where a vast number of "open source" reports have been produced purporting to "uncover" Cuba's secret biological weapons industry. (See, e.g., the Cuban American National Foundation (CANF) document summarizing these, "Bush's Campaign Against Bio-weapons Right to Include Castro's Cuba" [CANF, November 26, 2001], which includes the testimony by former Soviet Bioweapons expert, Ken Alibek ["Biohazard"].) The sorts of anxieties that fueled Bolton's speech can be more readily mined from the Senate hearing, detailed in "Cuba's Pursuit of Biological Weapons: Fact or Fiction?" (June 5, 2002) and available at http://www.access.gpo.gov/congress/senate.

33. Alibek, "Biohazard."
34. Nelson Valdés. "Fidel Castro, Bioterrorism and the Elusive Quote," *Counterpunch* (May 28, 2002).
35. See, e.g., Reuters, June 24, 1999, for claims that Cuba was producing biological weapons made by the *Miami Herald* on June 20, 1999, the same week that Limonta was removed; 2001 was not the first time these two sets of claims emerged in conjunction with one another.
36. Fidel Castro, *Orbe* (May 11–17, 2002): 5. This was accompanied by a declaration to the same effect by the Ministry of Foreign Relations in the popular magazine *Bohemia* (*Bohemia* 6 (1994): 27).
37. "Quines son las Verdaderas terroristas, IV," *Juventud Rebelde*, tabloide especial, May 14, 2002.
38. These were published in full in *Granma*, May 13, 2002.
39. This figure, released on state television, is consistent at least with the figure of 100,000-plus given in *Granma* the following day.
40. See also "Carter: EE.UU. sin evidencias sobre bioterrorismo," *Granma*, May 13, 2002. The Carter Foundation had also carried out work in South Africa where Cuba was involved in trying to secure generic drug markets.
41. US State Department press release and on-camera interview, May 13, 2002.
42. Quoted in the *Boston Globe*, May 15, 2002.
43. See www.cubanet.org/CNews/y02/nov02/2702.htm.
44. Oxford English Dictionary.
45. Andrew Lakoff, *Pharmaceutical Reason* (Cambridge: Cambridge University Press, 2005), 174–77.
46. Douglas Starr, "The Cuban Biotech Revolution," *Wired* magazine 12, no. 12 (December 2004).
47. Kaushik Sunder Rajan, *Biocapital: The Transformation of Post-Genomic Society* (Durham, NC: Duke University Press, 2006).
48. Kiran Mazumdar-Shaw, interview with *NASDAQ India* entitled "Mighty Molecules," November 22, 2001.
49. "Biocon's Cuban JV Gets US Senate Approval for Vaccines," *The Financial Express* (July 27, 2004); see also YM Biosciences *Annual Report* (2005): 7.
50. Andrew Pollack, "U.S. Permits 3 Cancer Drugs from Cuba," *New York Times*, July 15, 2004.
51. Registration Statement for Securities Offered Pursuant to a Transaction, available online at Sec Info, http://www.secinfo.com/dsvRq.v3pb.htm. CancerVax would later go bankrupt.
52. I take this title from an article written by Zena Olijnyk in the online journal *Canadianbusiness.com*, "Here's to Us, Who's Like Us?" July 8, 2002.
53. José Ramiro Más Camacho, Lázaro Ramos Morales, and Sergio Pérez Talavera, "Dirección estrategica en el Centro de Ingeniería Genética y Biotecnología," presentation (n.d., but probably 1999).
54. Agustín Lage, "Las biotecnologías y la nueva economía," 61.
55. Cited in Pollack, "US Permits 3 Cancer Drugs from Cuba."
56. As it would turn out, it would also depend on whether being "first in class" or "best in class" was the more important.
57. "President Lauds Biocon's Cancer Drug," *Times of India*, May 13, 2006.
58. Roy Porter, "Offering Resistance," *New York Times Review of Books*, June 29, 1997.
59. Nikolas Rose, "The Politics of Life Itself," *Theory, Culture & Society* 18, no. 6 (2001):

1–30; see also Nikolas Rose, *The Politics of Life Itself: Biomedicine, Power, and Subjectivity in the Twenty-First Century* (Princeton, NJ: Princeton University Press, 2006), and Gerry Kearns and S. M. Reid-Henry, "Vital Geographies: life, luck, and the human condition, *Annals of the Association of American Geographers* 99, no. 3 (2009): 554–74.

60. Shapin here is developing Niklas Luhmann's notion of "system trust"—that modern trusting is trust without familiarity and without effective possibility of mistrusting (Steven Shapin, *A Social History of Truth: Civility and Science in Seventeenth-Century England* (London: University of Chicago Press, 1994): 411.

CHAPTER EIGHT

1. The full speech "Science Matters" is available on the Downing Street website at http://www.number10.gov.uk/output/Page1715.asp.
2. John Gribben, interview, London, 2006.
3. Steven Shapin, *A Social History of Truth: Civility and Science in Seventeenth-Century England* (London: University of Chicago Press, 1994), 17.
4. Nancy Scheper-Hughes, "The Global Traffic in Organs," *Current Anthropology* 41, no. 2 (2000): 191–224, and Adriana Petryna and Arthur Kleinman, introduction to *Global Pharmaceuticals: Ethics, Markets, Practices*, ed. Adriana Petryna, Andrew Lakoff, and Arthur Kleinman (Durham, NC: Duke University Press, 2006).
5. Michel Foucault, introduction to Georges Canguilhem, *The Normal and the Pathological* (Zone Books: New York, 1989), 21.
6. Jon Agar, Crosbie Smith, and Gerald Schmidt, eds., *Making Space for Science: Territorial Themes in the Shaping of Knowledge* (London: Palgrave Macmillan, 1998), 2.
7. Gaston Bachelard, *Le materialisme rationnel* (Paris: Presses Universitaires France, 1953), 86.
8. Louis Althusser, *For Marx* (New York: Pantheon Books, 1969). For a useful discussion of this in precisely the context of pharmaceutical production, see Kaushik Sunder Rajan, *Biocapital: The Transformation of Post-Genomic Society* (Durham, NC: Duke University Press, 2006), 6.
9. An editorial in *Nature* 436, no. 7049 (July 2005) declared that Cuba had developed perhaps the most significant scientific research capacity outside of Southeast Asia. "It is worth asking how Cuba did it," the editorial continued, "and what lessons other countries might draw from it."
10. See Agustín Lage, "On the Cross-Fertilization between Biotechnology and Immunology: Current Situation in Cuba," *Vaccine* 24, no. 12, suppl. 2 (2006): S2–6.

# INDEX

Acquired Immune Deficiency Syndrome (AIDS), 92, 94, 142, 154
Albright, Madeleine, 140
Alibek, Ken, 159, 192, 193
Allan, David: on company strategy, 103, 106; on Cubans' perspective, 150; on framing Cuban biotech, 110–12; on embargo, 124, 145; on prospects for integration, 141, 143; on prospects for products, 158
Althusser, Louis, 168
America. *See* United States
American Cancer Society, 16
applied versus basic research, 82–83
Argentina, 4, 152
Asilomar, 19, 20
*Avances Medicos de Cuba*, 145

Bachelard, Gaston, 167
Barrueco, José Miyar (Chomy), 37, 58, 131
Beta Gran Caribe, 116, 142
BIOCEN, 141
Biological Front, 48, 49, 51, 52, 60, 76, 131
biological weapons: claims of, 147; response to, 150, 152
biomedicine, 39, 82, 104–5
Biomundi, 91, 132
biosecurity, 129, 134
biotechnology, Cuban, 22, 35; cancer, work on, 94–102, 112–14; emergence of, 163; foreign capital, need for, 107, 142, 150; funding, 82; immunotherapeutic approaches in, 99–100; interest of foreign companies in, 103, 106, 115 143–57; interferon, initial work on, 5, 13–22, 32, 41, 47–48; originality of, 165–68; pipeline of, 169; political clampdown in, 150; professionalization of, 109; public health roots of, 29, 31, 33–38, 108; representational framings of, 145–47; resistance to capitalist norms, 91–93, 107, 109, 110, 134, 137, 145, 157; shifting rationality of, 109–10, 121–23; training of personnel, 71–72; vaccine research of, 53, 63–67, 105, 124, 146, 155; work ethic in, 74
Biotechnology Industry Organization (BIO), 4
*Biotecnología Aplicada*, 144
Blair, Tony, 161
*Bohemia*, 1
Bolton, John, 138, 159
Borroto, Carlos, 21, 89, 154
boundary work, 129
Bravo, Ernesto, 78, 104
Brazil, 4, 65, 142, 163
*Breast Cancer Research and Treatment Journal*, 99
Britto, José Fernandez, 71
Burch, Kevin, 127
burdens of proof, 138, 148, 152, 159

Campa, Concepción, 59, 131
cancer research: in Cuba, 94–102; EGF and, 99–100; Cuban products from, 142, 155–59; history of, 94–95;

cancer research (*continued*)
    immunotherapy and, 96–97, 112, 158;
    interdisciplinarity and, 112; monoclonal antibodies and, 97–99; self/nonself tolerance, 100
Canguilhem, Georges, 39, 161; the normal and the pathological, 100, 166; critique of Marx, 163; and scientific ideologies, 164
Cantell, Kari, 15–16, 18, 43, 46–47
capitalism: and biotechnology, 4, 90, 163; flexible production, 69; and interstate system, 127; practices of, in Cuba, 64, 69; regional forms, 68–70; response to pressures of, in Cuba, 91–93, 107–9, 110, 134, 137, 145, 153–54, 157; venture capital, 4, 116; as way of life, 4
Caribbean basin strategy, 48
Carter, Jimmy, 151
Castro, Fidel: and collapse of Soviet Union, 89; on Cuban science, 25, 27, 52; on innovation, 68; on plans for biotech, 1–2, 6, 14, 35, 52; and medical metaphors, 146; response to bioweapons allegations, 147–51; visit to South Africa, 142
Center for Genetic Engineering and Biotechnology (CIGB), 1, 21, 41, 53–57, 80, 92, 101
Center for Molecular Immunology (CIM), 64, 94, 96–99, 111, 114, 142, 155
China, 4, 93, 141, 143
CIMAB, 111, 122–23, 126, 155–56
citizenship, 3–38, 84, 111, 147
Clark, Randolph Lee, 14, 18–19, 31, 158
clinical trials, 67; and CENCEC, 105, 130; in Cuba, 18, 52, 71, 100, 103–5; of Cuban products in Canada and Europe, 109, 148; ethics and, 104; as form of warrant, 112, 157, 165; licenses for, in United States, 141
*consagración. See* work ethics
Coronil, Fernando, 25, 26
Costa Rica, 4
Council for Mutual Economic Assistance (CMEA), 30, 55, 89
Council of State, 58, 64, 106, 131
Cuba: and clinical trials, 18, 52, 71, 100, 104; and colonial period, 24–25; intellectual property, approach to, 117–26; and history of public health, 33–34; and neocolonial period, 23, 25; pharmaceutical companies and, 115–17, 154–59; Rectification, 42; and science policy, 31–32; Special Period in, 89
*Cuba Business*, 141
Cuban Academy of Sciences (ACC), 32, 79, 109, 131
Cuban Communist Party (PCC), 31; and congresses of, 48, 108
Cuban National Office of Inventions and Trademarks (ONIITEM), 119, 121, 131
Cuban Office of Intellectual Property (OCPI), 131

Darwinism, 8
de la Fuente, José, 46, 91, 135, 139, 147, 148–49, 159
Delaporte, François, 19, 113
Delgado, Gregorio, 23, 36
dengue fever, hemorrhagic, 16, 18, 65–66
De Tocqueville, Alexis, 3, 7
development, 23–36; and science, 27; theories of underdevelopment, 25; nationalist and socialist visions of, 38, 41, 55; questions of in Cuba, 49; and biotech, 53, 90
Dewey, John, 28, 86
doability paradigm, 95, 100–101, 105, 113, 163

*Economist*, 112
embargo, US, on Cuba, 38, 123–28; constraints on Cuban marketing and sourcing, 125; exemptions to, 140, 156; impact on innovation in Cuba, 123, 135; impact on public health in Cuba, 38–39; intersection with TRIPS, 123–24
epidemiology, 113
epistemology: epistemic distance, 100, 105, 113, 162; epistemic geography, 166; epistemic space, 87; and political economy, 163–65; practical, 85; spatialized, 9, 167
European Medicines Agency (EMEA), 130, 157

Fernandez, Damian, 28, 34
Fernández, Luis Enrique, 67, 123

*Financial Times*, 140
Finland, 32, 43, 46
Finlay, Carlos J., 24, 113
Finlay Institute, 63, 64, 65–67, 101, 133
Food and Drug Administration (FDA), 130, 137, 140, 153, 157
Foreign Investment Law (1995), 107
Foucault, Michel, 22, 102; the author-function, 126; on Canguilhem, 166; and epistemic geography, 166; genealogical approach, 22
Frasch, Carl, 66, 140
Fujimura, Joan, 95, 122

Gambetta, Diego, 76
genetic engineering, 45–47
GlaxoSmithKline. *See* SmithKline
*Granma*, 1, 52, 91, 151, 152
Granovetter, Mark, 76
Gresser, Ion, 13
Guevara, Ernesto "Che," 27; as doctor, 37; influence of "moral man" debates on science, 85; "New Man" ideal, 36
Gutiérrez, Carlos, 107

Hacking, Ian, 22, 137
Havana, 57–58
Heber Biotec, 116, 126, 128, 139, 146, 153
Helms-Burton Act, 124
Herrera, Luis, 5, 47–48, 131, 151
historiography: and science, 167; critiques of, 22
hospitals: Calixto García, 33; Frank País, 76; Hermanos Ameijeiras, 103–4; M. D. Anderson, 14
Houston, TX, 15
Human Immunodeficiency Virus (HIV), 92, 94, 142, 154

India, 4, 25, 43–44, 45, 128, 155–58
informal versus formal mode of work, 51, 58–64, 75–77, 86–87, 141, 162, 181
innovation: capacity for, 85; and community, 75; in Cuba, 27, 30, 32, 65, 68, 78, 82–83, 120, 125; and geography, 68–70, 77, 86; and improvisation, 78; incentives for, 119; in science, 69; spatial innovations, 167–68
Institute for Animal Health, 7

Instituto Finlay. *See* Finlay Institute
intellectual property: in Cuba, 120, 145; cultural differences in, 118, 119–20; globalization of, 118–19; and innovation, 119; logic of, 117–18; in relation to sovereignty and territory, 126–27
interferon, 5, 13–22, 32, 41, 47–48
International Bank for Reconstruction and Development (IBRD), 25
International Center for Genetic Engineering and Biotechnology (ICGEB), 43–45, 52, 60
International Standards Organization (ISO), 132–34, 136, 141
Isaacs, Aleck, 13

Japan, 57, 69
*Journal of the National Cancer Institute*, 143
Juan Antonia Echeverría Scientific Polytechnic Institute (ISPJAE), 58
Junta Central de Planificación (JUCEPLAN), 30, 90

Kapcia, Antoni, 49
Kennedy, John F., 38
Kenya, 4
knowledge: and applied versus basic research, 83; biomedical, 84, 105, 155; geography of, 8–11, 167; embodiment of, 41; knowledge-making practices, 64; knowledge transfer, 43; and power, 136; and sanction, forms of, 124–26; specialist knowledge, 73; and value, 126; and warrant, 105
Kuhn, Thomas, 7, 85

labor: conditioning of, 74, 75; conditions for, 90, 135, 167; formation of, 33, 70; mobility of, 71; localization of, 76
Lage, Agustín, 59, 89, 96, 101, 140, 144, 151, 158–59
Lakoff, Andrew, 152
Laksey Partners, 117
Latin American School of Medicine, 151
Latour, Bruno, 8, 132
Lefebvre, Henri, 7
Lenoir, Timothy, 59
Limonta, Manuel, 15–16, 54–56, 60, 106, 139

Lindenmann, Jean, 13
liquidity, 135
Literacy Campaign, 47
Livingstone, David, 8
López-Saura, Pedro, 13, 15, 42, 43–47, 60, 80–81
Luhmann, Niklas, 159
Lysenko, Trofim, 26; and Lysenkoism, 27, 74, 149

Marcuse, Herbert, 84
Martí, José, 24
Marxism-Leninism, 49
Mazumdar-Shaw, Kiran, 155–56
Meers, John, 116–17
Mella, Carlos, 76, 126, 128
meningitis B, 65–67
Merton, Robert, 64
Ministry of Public Health (MINSAP), 31, 65
Ministry of Science, Technology, and the Environment (CITMA), 50
Moncada, Salvador, 125
Morocco, 4
Mukerji, Chandra, 59

National Assembly of People's Power, 52
National Center for Clinical Trials (CENCEC), 105, 130
National Center for Medicine Control (CECMED), 105, 130
National Center for Scientific Research (CENIC), 31, 50–51, 146
National Institute for Oncology and Radiology (INOR), 94, 97
nationalism, 28; nationalist discourse, 24, 174; and pragmatism, 35, 69; and socialism, 28, 38, 59–60, 164
*Nature*, 124
*Nature Biotechnology*, 148
Negrín, Sonia, 80, 151
neoliberalism, 9, 88, 105, 126, 157, 162–63
*New York Times*, 156
Nietzsche, Friedrich, 113
norms: diagnostic, 152; and ethics, 9; global, 11, 91, 130, 132, 134, 136, 153, 167; intellectual property rights, 118, 127, 132; regulatory, 4, 11; and standards, 133–36
North American–Cuban Scientific Exchange Program (NACSEX), 79
Norway, vaccine research in, 66–67
nuclear research, 107
Nuñez-Jímenez, Antonio, 27

Office of Foreign Assets Control (OFAC), 140, 153, 156
Office of International Collaboration (DCI), 107
Ortiz, Fernando, 2

Pan-American Health Organization (PAHO), 1
paradigms: doability paradigm, 95, 100; paradigm shifts, 84; regulatory, 103; research, 103; slow paradigms, 19–20
Pasteur Institute (Paris), 47, 94
Patent Cooperation Treaty (PCT), 121
patents: challenge of, 122–26; expiration of, 103; patent offices and granting of patents, 141; use of in Cuba, 145, 154. *See also* intellectual property
Pedro Kourí Institute (IPK), 115
Pérez, Rolando, 67, 70–71, 125
*Periodo Especial. See* Special Period in Time of Peace
pharmaceutical industry: Biocon, 154–56, 158, 161; Biognosis, 116–17, 139; Biovation, 143; CancerVax, 156–57; GlaxoSmithKline, 115, 116, 140–43, 146; and global standards, 117–19; Merck, 115, 124; Monsanto, 143; neoliberal pharmaceutical regime, 9, 91, 105; new formations of, 154–55; partnering in, 159; pharmaceutical lobby, 8; regulation of pharmaceutical products, 130; vaccines as part of, 169
Plan Programática, 38
Polo Científico del Oeste. *See* Science Pole
Porter, Roy, 159
Powell, Colin, 151, 152
PPG, 93
Programa Integral del Progreso Científico Técnico (PIPCT), 55

Puerto Rico, 124–25
Putnam, Robert, 76

Rabinow, Paul, 19, 77, 136
rationality: conflict between different rationalities, 110; and institutional embedding, 129–31; localized forms of, 161–62; political, 87, 90 (*see also* Special Period in Time of Peace); scientific, 2
Reagan, Ronald, 48
Rectification of Past Errors and Negative Tendencies, 42, 48–50, 53, 58–60, 90
regional innovation systems, 68–70
regulation: in biotechnology, 126; of expertise, 85, 103, 136; and intellectual property, 127; and overregulation, 136–37; of production processes, 129–30, 134
research and development (R&D), 66
Rhoads, Cornelius, 95
risk-taking, 77, 86, 93, 102, 106, 109–10, 134
Rogés, Germán, 115, 117
Rose, Nikolas, 159
Royal Society (London), 161

Santería, 6
Sasson, Albert, 56–57, 60
Saxenian, AnnaLee, 68
science: capitalization of, 90; and geography, 9–11; sanctioning of, 163; scientific norms, 132–34; 139; as vocation, 5; and warrant for, 105
*Science*, 112, 124
*Science and Policy* journal, 75
Science Pole, 57–59, 60, 85–88
Scientific Council, 66, 76, 80, 129, 131
scientific reason, 7, 9–10, 132, 134; and doubt, 152; Western and non-Western forms, 136
*Scripp*, 112
Shapin, Steven, 73, 159
Sierra, Gustavo, 63–65
Silicon Valley, 68–69, 77
Silva, Eduardo, 129
Simeon, Rosa Elena, 59, 121, 131
Sloan Kettering Research Institute, 95, 98
SmithKline, 115, 116, 140, 143, 146
social capital, 76. *See also* Putnam, Robert

socialism: commitments to, 64; in Cuba, 26, 27–28, 88, 90; different forms of, 48; distrust of, 165; and entrepreneurialism, 88; and ideology, 163; and postsocialism, 10; and rationality, 126; and science, 26; "socialist" drugs, 153; and socialist internationalism, 38; socialist trade and political relations, 45, 55, 123 (*see also* Council for Mutual Economic Assistance [CMEA]); and state formation, 26–29; and visions of development, 41–42
socialist humanism, 35, 36, 60, 69, 83
*socialismo*, 63, 70
South Africa, 141
sovereignty, 125; and territoriality, 126,
Soviet Union, 23, 48; glasnost and perestroika, 48; and Lysenkoism, 74; researchers in, 70
space: as constitutive element of scientific culture, 86; role of in innovation, 68–69; and regulation over, 126–28; space of operations, 87; as territory, 128
Special Period in Time of Peace, 89
standardization, 8, 94, 117, 119, 132
Starr, Douglas, 153
State Department, U.S., 10, 138, 140
sugar, 24–25
Sunder Rajan, Kaushik, 155
symbolic capital, 35

technology: antibody, 143; diagnostic, 100; ELISA, 97; open-sourced, 128; recombinant DNA, 95; therapeutic, 100; transfer of, 55, 128, 159
Thomas, Hugh, 30
*Time*, 112, 140
Tormo, Blanca, 122–23, 137
Trade Related Aspects of Intellectual Property (TRIPS), 8, 118–19, 126
Traweek, Sharon, 69
Trudeau, Pierre, 106
trust: bearing upon Cuban practice, 159–60; between Cuba and foreign partners, 106, 117, 149; forms of, in Cuba, 76; geography of, 114, 149, 160; importance in biotech, 138; lack of

trust (*continued*)
    trust, 165; problems of, 73; "system trust," 159

Ubre Blanca, 6–7
Ultramicroanalytic System (SUMA), 97
United Kingdom: bioscience companies from, 116, 143; business interest in Cuba, 116, 139; research parks in, 70; science in, 161
United Nations Industrial Development Organization (UNIDO), 1, 42–45
United States: admission of Cuban drugs to, 140, 156; allegations of bioweapons against Cuba, 150–52; cancer research in, 95; intellectual property and, 119; as market, 116, 25; relations with Cuba, 23, 38; scientific culture, 5–6, 8
USSR. *See* Soviet Union

vaccines, 65, 146, 169
value: capitalization of, 126; and geographical variation in, 128; and intellectual property, 127; and processes of revaluing, 129, 136
VA-MENGOC-BC, 5, 65, 106

Waksal, Sam, 137–38
*Wall Street Journal*, 65
Walter Reed Commission, 24
warrant: and clinical trials, 112, 157, 165; cultural overdeterminations of, 165; economic warrant, 102; and knowledge, 105; warrant-making mechanisms, 102–5
Weber, Max, 5, 58, 61
Western Havana Scientific Pole. *See* Science Pole
Wittgenstein, Ludwig, 137
work ethics, 74–75, 78
World Health Organization (WHO), 1, 96, 118. *See also* intellectual property
World Intellectual Property Organization (WIPO), 65, 121
World Trade Organization (WTO), 118–19

YM Biosciences, 103, 107–8, 110–11, 114, 157–58
York Medical, 103, 107–8, 110–11, 114. *See also* YM Biosciences